MARTY'S TOP 10

DIET & FITNESS STRATEGIES

IMPROVE THE QUALITY OF YOUR LIFE

Marty Copeland's HIGHER FITNESS

MARTY COPELAND

Marty's Top Ten

Marty Copeland's HIGHER FITNESS

Diet & Fitness Strategies

Marty's Top Ten Diet and Fitness Strategies

ISBN 978-0-9797047-0-3 30-7775

12 11 10 09 08 6 5 4 3 2

For more information about health and fitness, or prayer support and weight loss, visit www.martycopeland.com.

I would like to dedicate
this book to my beautiful
daughter Courtney.

"Courtney, it is because of your willingness
to give that I was able to write this book.
Your help with Jonathan was invaluable in
allowing me to complete many thoughts and
ideas. You have been patient, kind, and always
faithful to pray for me. It is hard to express the
love and appreciation I feel for you
in helping this dream become a reality.
Thank you, Courtney.
You are such an important part of who I am.

I love you, Mom."

Marty Copeland's
HIGHER
FITNESS

CONTENTS

Marty Copeland's HIGHER FITNESS

INTRODUCTION

ATTAINING HIGHER FITNESS

NO DOUBT ABOUT IT, we are very complex. With different pasts determining our outlook on life, different personalities and body types, we all face a variety of challenges. It is my experience from years of counseling that most of our physical challenges are quite similar, yet it is our spiritual and emotional perspectives that establish our capacity to overcome them. Whether our primary need for improvement is spiritual, emotional, physical, or all three, it is our ability to balance our progress that will determine our outcome. **Higher Fitness** is a progressive program designed to get you from where you are today to where you want to be. Whether you are in bondage to dieting, or whether you simply want to improve the quality of your life, **Higher Fitness** will propel you toward victory.

The first step is to take an honest look at your life, your health, and your body. Then, ask yourself these questions: Am I at peace with where I am? Am I fighting the same battles year after year? Have past failures made it hard for me to try again, or have I simply given up? **Higher Fitness** will change the way you see yourself and your potential to improve. While none of us are perfect, God has made a way for us to consistently get better. God not only desires to help you reach your fitness goals, but He has a plan that will elevate you from a lifestyle of dieting to a lifestyle of freedom and self-control.

I failed for many years to successfully lose weight and keep it off. There was a time when I could not imagine what it would be like to live life free from guilt, shame, and bondage. Now I control what, when, and how much I eat, and I am in the best shape of my life. It was the transforming power of God's Word that turned my life around. Not a day goes by that I don't thank God for my freedom.

Higher Fitness is not about going from ground zero to perfection. This program is about starting where you are right now and attaining higher. God wants our lives to perpetually grow and improve. Just as our physical bodies grow from infancy to adulthood, so should our growth in spiritual and mental maturity be evident as well. Spiritual maturity is a part of God's plan for us and is vital to living a fulfilled life.

By increasing our spiritual and mental strength, we are able to implement the steps necessary to gradually change our bodies and effectively accomplish our goals. All of us must find our own balance between spiritual growth and practical application. The obese person who contributes to family and community yet finds himself at the age of forty on the brink of physical death is just as out of balance as the person who exercises incessantly or starves herself to meet society's criteria for beauty. **Higher Fitness** will not only inspire you to achieve a healthier balance in your life. It will also empower you to live a higher quality of life. Now, let's go higher.

10 "DO'S" TO BETTER NUTRITION

CHAPTER 1

Marty Copeland's HIGHER FITNESS

ONE

10 "DO'S" TO BETTER NUTRITION

WHAT WOULD THE PERFECT DIET BE? Well, let's see. In a perfect world we would have plenty of fertile soil. We would eat (and like) home-grown organic fruits and vegetables. We would have natural, farm-raised chicken, beef, and, of course, fresh fish. And we would drink 100 percent pure water. The problem is that the soil is not so great. Consequently, many foods are depleted of vitamins and minerals. In today's world, it's getting more and more challenging to find quality tap water. Most of us don't have access to good quality organic fruits and vegetables, and many of the animal products that we buy are not "natural" or lean. Even when we try our very hardest, it takes so much time and work to find and prepare quality foods that it can be downright frustrating. We're in a hurry, and who has time to plan that nutritious meal? It is much more convenient to drive through McDonald's and forget the repercussions. We are a fast-food society, and there is one thing about fast food—it's fast.

While there are a million "do's" and "don'ts" where nutrition is concerned, I'm going to give you 10 "do's" that will move you closer to where you want to be. I've learned that in doing the "do's," many "don'ts" will automatically take care of themselves. It is much more effective to keep our focus on what we should do instead of always thinking about what we shouldn't do. So don't be discouraged if you're a junk-food junkie, because these ideas can easily be implemented into your diet. If you replace a Coke with a piece of fruit today, you are one step closer to higher fitness.

Remember to focus more on the foods that you must eat to increase your health, and not so much on what you shouldn't have.

1. **Do add at least 1 - 2 cups of vegetables into your daily diet.**

   Studies show that we need about 9 servings of vegetables and fruits each day to nourish our bodies and help prevent disease. One vegetable serving is 1/2 cup chopped raw vegetables, 1/2 cup cooked vegetables, or 1 cup raw leafy vegetables. Your first step is simply to up your vegetable intake by 1 to 2 cups each day. (No, french fries and ketchup don't count.) One of the easiest ways to get 2 cups of vegetables, is simply to have a salad. Make an effort to make your salad with several raw veggies and add your lettuce last. Try dipping your fork into your dressing to avoid consuming too many calories.

2. **Do eat at least 1 piece of fresh fruit every day.** If you are already eating 1, then increase it to 2. It's more fun to eat healthy if you give your taste buds a little variety. Have a banana and strawberries one day, and an apple and orange the next. Or make a fruit salad on Sunday night to last you all week. Because of the

amount of natural sugar in fruit, it is a good idea to have your fruit by midafternoon if your goal right now is weight loss.

3. **Do increase your water intake**. Headaches are sometimes brought on by insufficient water intake. You simply cannot be healthy without drinking enough water. Water helps cleanse your body and is necessary for fat loss. Water can help your complexion and even boost your mood. Your goal would be about 64 ounces of water for the average-sized person, more if you are bigger or when engaging in physical exercise. Remember, progressive change is the name of the game. If you haven't been drinking any water, start with 2 glasses a day. If you've been drinking 2 glasses, increase to 4, and so on.

   Sometimes it's very difficult to find time to fill up and drink 8 glasses of water. Here's a tip that really makes it easier for me. I keep a couple of 33-ounce water bottles in the fridge. Somehow it seems really easy to guzzle down 2 of those a day. It's also a good idea to keep a water bottle in the car, next to your computer at work, in the kitchen, on your nightstand, and anywhere else you find yourself throughout the day. That way, you'll have a water bottle handy all day as you sip your way to 64 ounces. Simply refill your bottles at your kitchen sink, but remember, the purer your water, the better. Purchasing a purification system such as reverse osmosis is an excellent idea.

4. **Do add a high-fiber cereal to your diet.** If cereal is your breakfast of choice, keep the excellent health benefits of fiber in mind. According to the USDA Beltsville Human Nutrition Research Center in Maryland, increasing your fiber intake will help you digest about 130 fewer calories per day. (A 1/2 cup of *All Bran* cereal has 10 grams of fiber.) If you don't like the high-fiber cereal choices on the market, then simply mix a little bit of a high-fiber variety with your regular cereal. You can also add berries to your cereal for a more healthful breakfast. Blueberries are chock full of antioxidants, those cancer fighting agents.

5. **Do try a higher protein breakfast a couple of days a week.** It's great to eat some lean protein at every meal. When I eat protein with breakfast, I don't get as hungry as quickly. Make Tuesdays and Thursdays omelet days. Mix a couple of eggs with peppers, onions, and mushrooms. If you regularly eat a lot of eggs, try 1 whole egg and 3 or 4 egg whites. This will cut your cholesterol and still taste great! For that extra splash of taste add some picante sauce. (This is making me hungry.)

6. **Do add whole grains to your diet.** The best choices are at the health food store, but there are some decent ones at the supermarket. Whole grain breads are so much more nutritious than enriched wheat or white. I have white and enriched wheat bread on my "junk food" list and whole wheat and whole grain on my

"healthy" list. A turkey, chicken, or tuna sandwich with lettuce and tomato is much more satisfying on denser bread than on white bread. If you've never tried it before, sneak it into your diet gradually. Have a slice with your breakfast with a thin spread of all-fruit jelly or peanut butter.

7. **Do add fish to your family's diet.** Try baked, broiled, or grilled fish at least once a week. Salmon and orange roughy are my favorites. Salmon and some other cold-water fish contain EFAs (Essential Fatty Acids). These unsaturated fats are not produced by the body and are needed in a healthy diet to help prevent disease. I wrap the orange roughy in foil, season with Mrs. Dash garlic and herb, lemon juice, and cover it with onion and red, yellow, and green peppers. Sometimes I use different veggies. Bake and serve. Experiment with different fish recipes until you find some that your family will eat. It's a great alternative to chicken!

8. **Do plan ahead!** Reserve a piece of fruit, string cheese, or half a turkey sandwich and a bottle of water for your ride home from work or school. It is very important to stabilize your blood sugar level in the afternoon to avoid excessive eating in the evening. This seems to be the most difficult time of day for people—the time when they munch uncontrollably, consuming way too many calories and unhealthy snacks. So keep your blood sugar level stable by eating

something small and healthy every three or four hours—especially at peak hungry times. Never come home from a stressful day famished.

9. **Do use olive oil when cooking with fat.** Olive oil is an Omega 6 Essential Fatty Acid and is unsaturated. Of course, if your goal is fat loss, you will want to use your oil in smaller portions. Nonfat cooking sprays are also a good choice when olive oil is inappropriate.

10. **Do get the kids involved.** One night a week try making homemade pizza or something else they might enjoy. Be sure to have all their favorite ingredients and let them take turns creating their own recipes. Start with a regular or whole-wheat pizza crust, tomato sauce, part skim mozzarella cheese, and any favorite vegetables. If they like chicken or hamburger, simply let "pizza night" follow a previous meal of chicken or beef. Using leftovers is a great way to save time. Encourage them to have fun. When you perfect a pizza creation that everyone likes, be sure and write down the recipe so you won't forget it. This will be a great way to spend some family time together, and your kids will probably keep using these recipes when they leave home.

FACT. . .

A series of studies at Philadelphia's Monell Chemical Senses Center revealed that we acquire a taste for the foods that we are most often exposed to. If we regularly eat a lot of high-fat foods, then we will prefer them. According to researcher Richard Mattes, Ph.D., R.D., if we eat lower-fat foods for several months (the study lasted six months), we will "develop a heightened acceptance of low-fat foods over high-fat foods."[1]

10 WAYS TO ESCAPE TEMPTATION

CHAPTER 2

Marty Copeland's

HIGHER FITNESS

TWO

10 WAYS TO ESCAPE TEMPTATION

■

"There hath no temptation taken you but such that
is common to man: but God is faithful, who will not
suffer you to be tempted above that ye are able;
but will with the temptation also make a way of escape*,
that ye may be able to bear it."

1 CORINTHIANS 10:13 KJV

■

*TO ESCAPE MEANS, "TO GET AWAY,
TO AVOID A THREATENING EVIL."

AFTER MANY YEARS OF PROGRESSING in this area, I have learned that there are times when it is simply smarter to avoid temptation altogether. Common sense tells us that just as an alcoholic who quit drinking last Tuesday should not get a job as a bartender, neither should an overeater hang out with food. I'm not trying to minimize the difficulty that it might entail to quit drinking, but rather pointing out the obvious. We must eat to live! While there are many people challenged simply by making the right nutritional choices, the observation from my own experience with past overeating habits and with the habits of those I have counseled is that the *greatest* challenge most overeaters must face is simply to stop eating once they start. This is especially hard if your mother always told you "to clean your plate!" Whether our dieting/overeating lifestyles have nullified our hungry/full response, or whether we have become masters at ignoring that response altogether, we must take measures to address this issue wisely.

Even when temptation cannot be avoided, it can certainly be outsmarted. Until enough self-control to hang out with food at the end of a meal is developed, here are some preplanned ideas that are tested and tried. First, whether you are making your own dinner or eating off a buffet line, you must consider that you are preparing a plate for a friend who is hungry yet trying to be healthy. I have actually asked people who struggle in this area to pretend that I was coming over for dinner. Knowing that I was interested in healthy eating, yet not on a starvation diet, how would they prepare my plate? This simple exercise seemed to help them be more objective about their own portion sizes. The goal is to not let our appetite determine our portion size but rather our awareness of what we already know about health.

Second, you must take a few minutes to make a list of your top 10 "to do's." Each item on your list should involve an activity that requires getting your mind and your eyes off of food. That, of course, eliminates TV. Also, each "to do" should take about 10 to 20 minutes, although some may take longer. The activity should be productive but not overwhelming. Again, we are emphasizing the impact of small, progressive change.

Finally, you must make a quality decision before you start that at the moment of temptation you will pull from your "to do" list whenever it is possible to do so. This decision must supercede any feelings you might have about whether you want to continue eating or not. The decision has already been made that when you know that you have eaten enough food to satisfy hunger and nutritional needs, you will immediately "escape" your eating environment. Eliminate any feelings of deprivation by reminding yourself that you are not on a "diet" and you are free to eat again when you are hungry. You are simply training yourself to exercise control over food by taking action. You are developing the habit of self-control.

Because we are all at different levels and different seasons of our lives, you will want to custom design your own escape routes to accommodate your priorities and maximize your productivity. Needless to say, you will consistently need to update your list. Here are some ideas to help you get started.

1. Clean out and reorganize two dresser drawers or the top shelf of the hall closet.
2. Shoot a round of hoops with the kids or call that friend or relative you've been meaning to call for weeks.
3. Clean out and vacuum your car. If you have time, drive the car to the car wash.
4. Go through two stacks of the junk mail sitting on your desk.
5. Turn on some fun music and get the kids to each dust a piece of furniture with you.
6. Get the kids bathed, or if you don't have kids, give the dog a bath, or get a dog, or better yet, take a long, hot, relaxing bath without the kids or the dog.
7. Replace burnt-out light bulbs and then make a list of other household items you need from the store.

8. Throw a load of laundry in the washer and clean and reorganize the laundry room mess. Yes, I am assuming your laundry room is messy.
9. Go through one section of your closet and get rid of the stuff you never wear and put it in a bag to take to an outreach or have an organization such as Goodwill come and pick it up. Do another section of your closet next time and get rid of those extra hangers!
10. Go sit outside and think about your progress. You should quickly be noticing that you are more in control of your life. The point of taking on small tasks is because doing the whole job at one time is probably overwhelming, and that is why it has not already been done.

Please note that by using your "to do" list as a way of escaping temptation, you are accomplishing two very important goals at once. First, you are developing the habit of stopping your eating because you have decided to stop. You are developing this habit of control by choosing to get your mind off of food and taking the necessary action to do so. You will find your thinking process after eating begin to change, and this will continue to get easier.

Second, you are alleviating one of the major causes of stress in your life, which is too much to do in too little time. As your "to do" list consistently gets done, you will begin to enjoy your progress and your home more and more. Realizing how productive you can be in short quantities of time simply by being consistent will help rid the pressure of an "all or nothing" mentality.

Gaining control of your time and reducing stress is an important part of a balanced lifestyle.

IT'S A LITTLE KNOWN FACT...

You might not know that there is another great benefit from developing the habit of getting your mind off of food for a while. When you recognize that you have eaten enough yet still don't feel satisfied, it should encourage you to know that it takes 10-20 minutes for your stomach to tell your head that it's full. So if you want dessert, delay it! By the time you're done with your after-meal activity, you'll probably find that you've lost the craving!

10 QUICK AND EASY SNACK AND MEAL IDEAS

CHAPTER

3

Marty Copeland's
HIGHER
FITNESS

THREE

10 QUICK AND EASY SNACK AND MEAL IDEAS

WE JUST SAW HOW MAKING DECISIONS ahead of time gives us the upper hand in some of our most challenging areas. A little organized food preparation is another great weapon to use against both snack attacks and the "what's for dinner?" dilemma. Pick one or two evenings a week to plan menus and shopping lists. It is crucial to take the time to stock your refrigerator with healthy, ready-to-eat snacks. If you come home completely famished, you're probably not going to take time to chop up some raw veggies. And if there is no fresh fruit, you might be tempted to grab some chips or a cookie instead. Keeping healthy choices available is an extremely important part of the process of creating a healthier lifestyle. It's what you do consistently that will determine the outcome of your efforts.

To avoid getting bored, cut up some fresh cantaloupe, honeydew, blueberries, and strawberries one week, and alternate grapes, apples, pineapple, and watermelon the next. For an extra splash, top a fruit salad with a dab of strawberry yogurt. Try to keep sliced carrots, celery, and some red and green peppers ready for dipping. Roll up some turkey slices to add a little protein to your snacks. Here are some additional snacking suggestions to get you on the right track.

10 EASY SNACK FAVORITES

1. Apple slices and two teaspoons of peanut butter for delicious dipping.
2. One whole-wheat tortilla packed with turkey and low-fat cheese. Pop it in the microwave until cheese is melted.
3. 10 almonds. 1/2 frozen banana.
4. Raw carrots and celery sticks.
5. Pickles. I absolutely love munching on a huge dill pickle. Because of the sodium, I make it an occasional snack. It's also a great substitute for high-fat popcorn at the movies.
6. Apple slices and a strip of low-fat mozzarella string cheese.
7. 1/2 cup blueberries with low fat yogurt.
8. 1/2 cantaloupe, just eat it out of its natural bowl.
9. A handful of pretzels and raisins.
10. Baked chips with salsa.

I must confess that cooking is not one of my favorite activities to do. Actually, it's not just the cooking I find challenging, but the meal planning, grocery shopping, table setting, getting everyone to the table while keeping the food hot, and the kitchen cleaning that seems to take so much coordination. I realized that I was going to have to get more organized with my efforts when my husband asked me a few years ago, "Are you doing some kind of an experiment to see how long we can survive without going to the grocery store?" I took the hint.

While there is no doubt still much room for improvement, we are rarely without fresh fruit, healthy snacks, organic milk, and a few home-cooked meals a week. While my dream is to one day have my own "private chef" to prepare wonderful, creative, and healthy meals for my family (and to clean the kitchen afterward), I will continue to search for shortcuts to getting food on the table. Here are a few more quick ideas just to make life easier. If you love to cook, these ideas may not be creative enough for you and are certainly not to be compared to an elegant, home-cooked meal. They are, however, much better than a fast-food drive-through. So, if your spouse and kids will allow you this timesaver, pick one night a week and keep it simple!

TEN QUICK DINNER IDEAS

To maximize your healthy eating efforts, please add a side salad or raw veggies to all meals not already including several vegetables.

1. **Almost a Patty Melt**—This is a lower-fat, lower-calorie version of the "patty melt." Grill or cook some lean beef patties or, better yet, turkey burgers. Sauté onions and bell peppers in nonfat, butter-flavored cooking spray. Place burgers on dark rye toast and cover with onions and peppers and a little spicy mustard. If you don't mind a few extra calories, melt a thin piece of low or nonfat Swiss cheese over the meat. "Yum."

2. **BLTs**—My husband loves these. Spread some whole-wheat toast with safflower mayo, add turkey bacon, (regular only if you have to), lettuce, tomato, and, if you like, avocado. Serve with a small salad and balsamic vinaigrette dressing. The "Newman's Own" brand is my favorite.

3. **Pita Sandwiches**—Stuff some whole-wheat pita bread with leftover chicken, turkey, or tuna salad. Add plenty of cucumbers, lettuce, and tomatoes and douse with Italian or French dressing.

4. **Open-faced Turkey and Swiss**—Spread your favorite mustard on whole-wheat toast. Top with turkey and low-fat baby Swiss. Melt under the broiler, in the toaster oven, or, for the quicker but soggier version, in the microwave. Top with sprouts for that "California" look. Serve with raw baby carrots, celery sticks, and/or a few baked chips.

5. **Scrambled Egg Sandwiches**—This is sooo simple and one of my favorites. Scramble your eggs and add a little salt and pepper to taste. Spread a thin layer of safflower or low-fat mayo on whole-wheat toast. Serve with a side of turkey bacon and a small glass of orange juice.

6. **Tuna/Garden Salad**—Wash and drain some white tuna (for a less "fishy" taste) and mix with safflower mayo and dill or sweet pickle relish. Put a scoop of tuna salad atop a salad of all the vegetables you like. Serve your favorite dressing in a small dish on the side and dip

your fork into your dressing. Watch those really high-fat dressings!

7. **Reuben Wanna-Be**—Toast some dark rye bread and use with a spicy mustard spread. My favorite is honey sweetened and contains a little horseradish. Cover one piece of toast with lean corned beef and a thin slice of low-fat Swiss cheese. Place under broiler or in toaster oven until cheese is melted. Top with other bread and serve hot.

8. **Vegetable Omelet**—Sauté mushrooms, onions, green and red peppers, tomatoes, and whatever else you want in some nonfat cooking spray. While veggies are warming, mix your eggs and pour into skillet. You may also cook the veggies and eggs together for an even faster meal. You can add a little low-fat cheese or cooked turkey bacon if you like. Add the tomatoes last. Whether you fold or flip your eggs, make sure they are cooked in the center of the omelet. Do not overcook. I like mine with hot sauce and whole-wheat toast. The kids might prefer it with a little ketchup and without the onions.

9. **Turkey and Cheese Quesadillas with Salsa**—Fold a whole-wheat tortilla over some low-fat Mozzarella or Swiss cheese and turkey. Be sure a little cheese is on both sides of the turkey to make the tortilla stick shut. You can pop these in the microwave or place in a toaster oven for crispness. If you warm these in the toaster oven or under the broiler, spray the outside of

the tortilla with some "I Can't Believe It's Not Butter." This will turn it a golden brown. Serve with salsa, hot sauce, or low-calorie dressing.

10. **Whole-Wheat Submarine Sandwich**—Pretty simple, just focus on lots of fresh veggies and go for the leaner cuts of meat. I always prefer turkey. Add shredded lettuce, tomatoes, peppers, pickles, and even jalapeños. Skip the olives for right now if you are losing weight. Try using mustard instead of mayo or stick with a little safflower mayo. I like a squirt of low-fat ranch dressing across my sandwich.

IT'S A LITTLE KNOWN FACT . . .

It's good to snack. We're talking healthy snacks, of course. According to a study done at the University of Toronto, dividing the same amount of food that you would eat in three meals into five or six snacks and mini-meals can make you healthier. Those who normally ate three meals a day lowered their blood cholesterol levels and their risk of heart disease by snacking and eating smaller meals spread out throughout the day.[2]

10

EATING OUT TIPS

CHAPTER

4

Marty Copeland's
HIGHER
FITNESS

FOUR

10 EATING OUT TIPS

I CAN REMEMBER SO MANY TIMES when I would be trying to lose weight and would do "good" all day and then blow it in the evening, especially when we would go out to dinner. Either from not eating enough calories during the day, or from the smells and visual stimulation of dinner party buffets, my physical desire for food would overcome my mental desire not to overeat. No matter how much self-control you have, it is simply not wise to show up at a dinner party feeling as though you are starving.

1. **My favorite and most effective tip** for avoiding pitfalls at dinner is to drink a protein drink or eat half of a sandwich with a bottle of water 1/2 to 1 hour before dinner. This allows you to make intelligent decisions on what foods you should eat to help you reach your fitness goals. It will also help you identify to what degree your eating habits are affected by reasons other than physical hunger. Remember that taking the time and effort to avoid uncontrollable hunger is an excellent way to ward off an uncontrollable appetite. (Protein drink recipe at end of chapter.)
2. **Call ahead!** If you are at a level of spiritual development where you don't yet trust yourself to make the

right decision at the restaurant, call ahead and ask what healthy choices are on the menu. Many restaurants are happy to help you, and if you have a fax, they can even fax you a copy of their menu. So if you know your dining destination ahead of time, make your choices before you arrive. That way you won't be so tempted to order a high-fat entrée. You'll have a plan.

If you are going to a dinner party at a friend's house for dinner, simply call and tell them you are eating healthy and would like to know what they are serving. Done in a polite way, they should take no offense at your asking. If their menu is barbecued pork, fried chicken wings, and several desserts, simply eat a good meal right before you go. Never ask them to prepare something different for you. They are busy enough getting ready for the party. On the other hand, if you are entertaining, always try to have nutritious foods as part of the menu in honor of those who are conscious of their health.

3. **Pack a snack**. What if you are not at home, but are going from work or class directly out to dinner? Keep a small lunch box or ice chest at work or in the car with a bottle of water and a healthy snack to eat on your way. By the time you reach the restaurant, you will feel much more in control. Countless times before eating at a certain restaurant (especially Mexican restaurants, which are my favorites), I would look at the menu. And though I might see my favorite entrée,

because my snack had curbed my appetite, I would choose a salad or soup and side dish instead. I try to remind myself to make decisions ahead of time, based on my fitness goals. If maintaining my weight is my goal, I choose between the lighter meal just mentioned or go ahead and order my favorite entrée. I'm free to make that decision because by the time I get to a maintenance program, I have enough self-control not to overeat.

When my goal is losing weight, I have already decided that I will make the choices that will help me reach my goals faster. When I first heard of people carrying around food, it seemed a little extreme. However, when I began studying health and fitness, I saw that many people in the health industry do this whenever necessary. When I am maintaining my weight, I normally don't practice this, but when I am losing weight—I do whatever is necessary to lose the extra fat. After so many years of struggling and then attaining victory over this weight-loss issue, I feel I have now become somewhat of an authority on the subject. I know it is okay to have setbacks, but I also know that every decision affects my goals. Quite honestly, I so despise the whole "diet/weight loss" issue that I want to achieve my goals sooner rather than later.

4. **Drink water before your meal comes.** Whether cooking at home or arriving at a restaurant, it's a great

idea to drink water before you eat. Some experts say that drinking anything at mealtime interferes with your digestion. If not drinking at meals is "the goal," then I will have to gain a lot more understanding before I ever attain that one. It just doesn't seem logical that we are not supposed to drink with our meals. Drinking 2 glasses of water before a meal seems like a great idea for two reasons. First, it will help you consume your daily water quota, and, second, it will help fill you up. This is especially important when you are unable to have your pre-dinner snack.

5. **Go to the bathroom.** I have actually been so serious about not overeating and spoiling my efforts to lose weight that I have gone so far as to have a pre-dinner routine. When I show up at a restaurant so hungry that everything looks good, this is what I do. I sit down at the table, drink a glass of water, order a dinner salad as an appetizer (no cheese, dressing on the side) and my dinner selection, then, while everyone else is eating chips and hot sauce or bread and butter, I make a bathroom run. Even when I don't need to go, I'll go wash my hands, put on some lipstick, just take a few minutes so that I'm not staring at the chips and hot sauce. I think about my goals, and by the time I return to the table, my salad comes right out. Once I eat my salad, I'm not so hungry anymore.

6. **Be curious.** Ask how their chicken or fish is prepared. I asked a waitress at a popular steak house

about their chicken preparation. She told me that it was marinated in so much butter that I would be much better off ordering a steak instead. No wonder their chicken was so juicy! Hidden fat in foods can absolutely ruin your most honorable efforts to make the right choices. Of course, baked, broiled, or grilled is always better than fried.

7. **Try something new!** Just gotta have a baked potato? Great, but instead of ordering it with butter and sour cream, try seasoning it with Pace Picante Sauce. Taste buds not so spicy? When they are available, try pouring vegetables sautéed in olive oil and "I Can't Believe It's Not Butter" over your potato. It's really good and much less fattening. I have even gotten to where I love a potato covered with cottage cheese, salt, and pepper. Remember, you are not giving up your favorite foods, but you are, however, modifying your choices in order to lose and burn excess fat.

8. **Share, share, share!** Split your meal with a friend or your spouse and consume half the calories. Most restaurants serve portions that are far too big for one meal anyway. If the entrée is too small for two, order grilled veggies, a small salad, or soup. (Clear soups are usually much less fattening than creamy soups.)

9. **Beware the colas!** Nothing can hinder your progress as subtly as drinking colas. Think about the calories. Ordering water or unsweetened tea at your meal can save you about 140 empty calories. That's a total of 280

calories if you get a refill! It's these kinds of small decisions that will make all the difference. It is just not smart to work hard at making changes but continue one habit that can literally sabotage your efforts. If the thought of breaking the cola habit is just too much, simply decide to have one at special occasions.

10. It's dessert time! When I'm out to dinner, I often order a decaf coffee or hot tea while my dining companions are drooling over the dessert menu. Just having something while they are having dessert keeps me from feeling left out. If I simply must have something sweet, but don't want to hinder my weight loss, I will ask for a bowl of fresh fruit. Most restaurants can come up with some fruit. I must admit that most of the people I hang out with are a lot like me. They enjoy dessert but choose to have it only on occasion.

When I do have dessert I usually share one with my husband. To some it might seem a greater victory to simply never order dessert at all. But for me to have just one or two bites of a dessert and push the dish away is a complete celebration of my freedom. I lived my life for many years with an "all or nothing" relationship with food. I used to watch people eat a little piece of dessert, and it seemed so unthinkable that they could make themselves stop. After all, they weren't even on a diet. I thought it was feast or famine. There was no in between for me. The strength of self-control rules my decisions now. When I decide to eat something not so healthy, and there are times I most certainly do (usually something with chocolate), it's a decision based on my right to choose, not on a compulsion. It's wonderful to be free!

IT'S A LITTLE KNOWN FACT . . .

Have you noticed that it's not just fast-food restaurants that are in the "supersize" business? From "Big Gulps" to "Jumbo Popcorns," it seems that convenience stores, movies, and many restaurants are also using larger portion sizes to attract more customers and make more money. As studies show, it is without question more challenging to control the amount that you eat when faced with larger portions. So be smart. If you want to stay trim or reduce your size, don't "supersize."

MY FAVORITE PROTEIN DRINK RECIPE

Mix in blender:

- 8 oz. organic skim milk
- 6-8 ice cubes
- 1/2-1 banana
- 1 scoop Nature's Plus Spirutein Powder (I like the chocolate flavor.)
- 1 Tbsp. Barlean's greens
- Barlean's Flaxseed Oil

  1 Tbsp. for a weight-loss program

  2 Tbsp. for weight maintenance
- 2 squirts of Concentrace Trace Mineral Drops

  (The Barlean's products and mineral drops can be purchased at the health food store.)

FEEDING YOUR SPIRIT

5

Marty Copeland's

HIGHER FITNESS

FIVE

FEEDING YOUR SPIRIT

■

"For He satisfies the longing soul,
and fills the hungry soul with goodness."
PSALM 107:9, NKJV

■

INSIDE EVERY HEART AND SOUL is the hunger to experience God's unconditional love. This love cannot be earned. It is a free gift to be received, therefore glorifying the gift itself. God is love, and His love feeds our spirit, satisfies our soul, and strengthens our physical body. It is the life and power of His love that enable us to live a fulfilled life. Many times, when our hearts are hungry for this experience, we "feed" that hunger with a variety of substances. Whenever we are spiritually low on love, we simply have less to give, and our lives become much more challenging. We must always take the time to honor God's gift of love so that we can strengthen ourselves and better love those around us. We must take the necessary time to feed our spirits.

For spiritual strength to come, we must nourish our spirit consistently. Here are 10 spiritual snacks to chew on every day. Take 10 minutes to repeat these out loud to allow them to get down into your heart. This is a process that can take time. This is a process that can change the rest of your life. Please continue to update your list as you search for those scriptures that particularly speak to your heart. Let's start with this simple prayer: "Lord, help these scriptures to come alive to me and help me to greater understand the love that you have for me. In Jesus' name, Amen."

1. **Proverbs 4:20-22 (NIV)**—"My son, pay attention to what I say; listen closely to my words. Do not let them out of your sight, keep them within your heart; for they are life to those who find them and health to a man's whole body."
2. **Psalm 63:1-5 (NKJV)**—"O God, You *are* my God; early will I seek You; my soul thirsts for You; my flesh longs for You in a dry and thirsty land where there is no water. So I have looked for You in the sanctuary, to see Your power and Your glory. Because Your lovingkindness *is* better than life, my lips shall praise You. Thus I will bless You while I live; I will lift up my hands in Your name. My soul shall be satisfied as with marrow and fatness, and my mouth shall praise You with joyful lips."
3. **Deuteronomy 28:2-8 (KJV)**—"And all these blessings shall come on thee, and overtake thee, if thou shalt hearken unto the voice of the LORD thy God. Blessed shalt thou be in the city, and blessed shalt thou be in the field. Blessed shall be the fruit of thy body, and the fruit of thy ground, and the fruit of thy cattle, the increase of thy kine, and the flocks of thy sheep. Blessed shall be thy basket and thy store. Blessed shalt thou be when thou comest in, and blessed shalt thou be when thou goest out. The LORD shall cause thine enemies that rise up against thee to be smitten before thy face: they shall come out against thee one way, and flee before thee seven ways. The LORD shall command

the blessing upon thee in thy storehouses, and in all that thou settest thine hand unto; and he shall bless thee in the land which the LORD thy God giveth thee."

4. **Psalm 103:1-5 (KJV)**—"Bless the LORD, O my soul: and all that is within me, bless his holy name. Bless the LORD, O my soul, and forget not all his benefits: Who forgiveth all thine iniquities; who healeth all thy diseases; who redeemeth thy life from destruction; who crowneth thee with lovingkindness and tender mercies; who satisfieth thy mouth with good things; so that thy youth is renewed like the eagle's."

5. **Isaiah 55:10-11 (KJV)**—"For as the rain cometh down, and the snow from heaven, and returneth not thither, but watereth the earth, and maketh it bring forth and bud, that it may give seed to the sower, and bread to the eater: So shall my word be that goeth forth out of my mouth: it shall not return unto me void, but it shall accomplish that which I please, and it shall prosper in the thing whereto I sent it."

6. **1 John 4:7-12 (KJV)**—"Beloved, let us love one another: for love is of God; and every one that loveth is born of God, and knoweth God. He that loveth not knoweth not God; for God is love. In this was manifested the love of God toward us, because that God sent His only begotten Son into the world, that we might live through him. Herein is love, not that we loved God, but that He loved us, and sent his Son to be the propitiation for our sins. Beloved, if God so

loved us, we ought also to love one another. No man hath seen God at any time. If we love one another, God dwelleth in us, and his love is perfected in us."

7. **James 1:12 (KJV)**—"Blessed is the man that endureth temptation: for when he is tried, he shall receive the crown of life, which the Lord hath promised to them that love him."

8. **Psalm 91:1-11 (KJV)**—"He that dwelleth in the secret place of the most High shall abide under the shadow of the Almighty. I will say of the LORD, He is my refuge and my fortress: my God; in him will I trust. Surely he shall deliver thee from the snare of the fowler, and from the noisome pestilence. He shall cover thee with his feathers, and under his wings shalt thou trust: his trust shall be thy shield and buckler. Thou shalt not be afraid for the terror by night; nor for the arrow that flieth by day; nor for the pestilence that walketh in darkness; nor for the destruction that wasteth at noonday. A thousand shall fall at thy side, and ten thousand at thy right hand; but it shall not come nigh thee. Only with thine eyes shalt thou behold and see the reward of the wicked. Because thou hast made the LORD, which is my refuge, even the most High, thy habitation; there shall no evil befall thee, neither shall any plague come nigh thy dwelling, for he shall give his angels charge over thee, to keep thee in all thy ways."

9. **Proverbs 16:3 (AMP)**—"Roll your works upon the Lord [commit and trust them wholly to Him; He will cause your thoughts to become agreeable to His will, and] so shall your plans be established and succeed."
10. **Joshua 1:8 (AMP)**—"This Book of the Law shall not depart out of your mouth, but you shall meditate on it day and night, that you may observe and do according to all that is written in it. For then you shall make your way prosperous, and then you shall deal wisely and have good success."

IT'S A LITTLE KNOWN FACT...

To meditate means "to dwell on anything in thought; to contemplate; to study, etc." In Joshua 1:8, the word "meditate" comes from the Hebrew root word "hagah," which also means "to murmur, speak, and talk."[3] What do we get in return for pondering, studying, speaking, and applying God's Words to our lives? We get prosperity, wisdom, and good success. I don't know about you, but I'll take that trade-off any day. What would more prosperity, wisdom, and good success mean to you and to those you want to help? I can use all the prosperity, wisdom, and good success that God is offering.

I LOVE/ HATE TO EXERCISE!

6

Marty Copeland's

HIGHER FITNESS

SIX

I LOVE/HATE TO EXERCISE!

"Please remember to consult your doctor before beginning this or any exercise program."

THERE ARE MANY PEOPLE who don't like to exercise, and, believe me, I understand. Many people think that I love to exercise. The truth is that I sometimes enjoy it and sometimes despise it. I must say that through the years, how I feel about exercise has become less important than how it makes me feel. Because I maintain a high fitness level, I usually enjoy the challenge of a good run—but it's the way the run makes me feel that I really like. On days that I exercise in the morning, I feel better all day long. I also feel great about myself, because I know that exercise is so good for me, and I exerted the discipline necessary to make myself do it.

I could exercise regularly during my first pregnancy, but seven years later, after the birth of my second child, I was in the worst shape I had been in for many years. While I was in good shape before I got pregnant, I was unable to work out at all during my pregnancy. Not only did I gain a lot of fat, but I also lost a lot of muscle. Three months after my son was born I began exercising again. Although it had been about a year since I had done practically anything, my brain still thought I should be able to run 2 or 3 miles, but my body literally could not even walk half a mile. I was gasping for air in less than 10 minutes and had to stop. In a few weeks I increased my time to 15 minutes, then 20. The first time I jogged, I lasted about 1 minute. Every fat cell in my body was screaming at me. It was miserable! I hated every minute and every 3,500-calorie pound that I had to burn.

Finally, after my body began to attain a higher fitness level, exercising became easier, and it began to make me feel so much better. Even then, I can't say that I loved it, but I did love the way that it made me feel. I felt younger, healthier, and more energetic. I also loved fitting back into my pre-pregnancy clothes and getting back to my pre-pregnancy size. You see, I spent so much of my life losing and gaining weight that even losing weight from pregnancy was, well, let's just say, not something I looked forward to. And guess what happened? Just a few months after returning to my pre-pregnancy weight, and about nine months after my husband's vasectomy, I found out that I was pregnant again. Oh the joy!

Once again I was unable to do hardly any exercise at all during my pregnancy, but these aerobic workouts, the weight training workout in chapter 7, healthy eating, and a lot of hard work helped get me back to my pre-pregnancy weight. Surely, for the last time!

So, where do we start? Most experts agree that 3 to 5 aerobic sessions per week for a duration of at least 20 minutes at 60 to 85 percent of your age-specific maximum heart rate is a good place to start. Beginning exercisers would start lower in their target zone. Advanced exercisers will exercise at the higher end of their target zone.

- **What is aerobics?** Aerobic exercise is any activity requiring oxygen that uses large muscle groups, is rhythmic in nature, and can be maintained for a period of time. Done consistently, aerobic activity trains the heart, lungs, and cardiovascular system to process and deliver oxygen in a more efficient manner. Therefore, an aerobically fit person can work longer and harder during an exercise session than someone who is not. And the aerobically fit person will also have a much faster recovery time. Whether your goal is weight loss or higher fitness, the longer the duration of your exercise, the more calories you will burn. You should check with your doctor before beginning this or any exercise program. If you are overweight or out of shape, start with 10 to 20 minutes and move up. Also, keep track of your heart rate.

- To keep your heart rate in your target zone, follow this simple equation:

MHR (maximum heart rate) = 220 - (your age)

Target Heart Rate Zone = MHR x .60 to MHR x .85

EXAMPLE—age 40.

220-40 = 180

180 x .60 = 108 (*low end target zone*)

180 x .85 = 153 (*high end target zone*)

If you are age 40, your Target Heart Rate Zone would be 108-153.

- To measure your heart rate: take your pulse for a 10-second count and multiply the number of beats times 6.

To keep things more interesting, mix your workouts up a little. One day take a 30-minute walk during your lunch hour, and the next day visit a gym with a friend and try out an aerobics class or a stairclimber. Here are 10 examples of aerobic exercise for you to choose from:

1. Walking/hiking
2. Jogging/running
3. Cycling
4. Swimming
5. Dancing
6. Kick-boxing
7. Roller-blading
8. Cross-country skiing
9. Stair climbing
10. Elliptical cross-training

While some might enjoy shooting a round of hoops with the kids or even an hour of running on the treadmill, others find it all too challenging. Well, it is challenging. Anytime change is desired, work is required. Whether our challenge is getting an unfit body fit, or getting a fit body fitter, the benefits far outweigh the challenge. How we think about exercise will determine how faithful we are to do it. Understanding the value of exercise can positively affect our attitude toward it. While for some, exercise may never be fun, with the right attitude it most certainly is doable and can even become enjoyable.

10 BENEFITS OF AEROBIC EXERCISE

1. **Exercise increases your stamina and endurance.** This ought to excite you. If you find yourself exhausted at the end of the day—too tired to even enjoy your family—you need to make regular exercise a part of your life. Once it becomes a habit, you won't want to give it up. You'll feel too good!
2. **Exercise reduces stress.** Exercise activates your endorphins, causing a natural high and a sense of well-being.
3. **Exercise reduces your risk for heart disease and stroke.** Coronary heart disease is the leading killer in the United States, and nearly half of them are women.
4. **Exercise strengthens your immune system.** It actually increases your resistance to stress and illness.
5. **Exercise strengthens your bones and joints.** This is a very important benefit for women because strong bones and joints decrease a woman's chance for osteoporosis.
6. **Exercise decreases your appetite.** You won't be as hungry because exercise acts as a natural appetite suppressant.
7. **Exercise will increase the number of calories you burn.** This increase will accelerate your weight loss and encourage weight control.
8. **Exercise will strengthen your heart and lungs so that they will work more efficiently.** Exercise

literally strengthens your heart muscle, and a stronger heart will last longer.

9. **Exercise decreases a woman's risk for breast cancer.** Studies have shown that women who engage in exercise at least three times a week have a lessened chance of developing breast cancer.

10. **Exercise decreases your blood pressure and blood cholesterol.** Exercise decreases your blood pressure and LDL, or bad cholesterol, while raising your HDL, or good cholesterol.

A consistent aerobic workout, coupled with weight training and healthy eating, will get us to our goals.

IT'S A LITTLE KNOWN FACT...

Can aerobic exercise make you smarter? *In his book,* The Aerobics Program for Total Well-Being, *Kenneth Cooper states that aerobic exercise can improve your intellectual capacity and increase your productivity. He conducted a study at Maxwell Air Force Base on a young group of lieutenants and found some interesting correlations. He found that those who performed best on a 12-minute aerobic endurance test, those who had the greatest endurance capacity, also made the best grades.* [4] *Even brisk walking 3 times a week for 45 minutes improves circulation and supplies more oxygen to the brain. More oxygen to the brain means an "increased ability to think and reason."*[5]

7

MUSCLE IS A MUST!

Marty Copeland's HIGHER FITNESS

SEVEN

MUSCLE IS A MUST!

■

WHY? BECAUSE A LEAN, TONED BODY LOOKS GREAT? Well, many people think so, but that is a matter of opinion. Muscle is valuable for many reasons, including its ability to burn calories. In fact, 1 pound of muscle burns about 35 calories a day, whereas 1 pound of fat burns only 2 calories.

Studies tell us that between the ages of 20 and 30, without weight-resistant exercise, we begin to lose muscle. As we age, the rate at which we lose seems to increase slightly. As we lose muscle, our basal metabolic rate slows down, which means we are burning fewer calories. This change in metabolism generally means a gain in fat. Let's say that between the ages of 30 and 40, we have lost 10 pounds of muscle and gained about 10 pounds of fat. That means that in 10 years, we would have a 20-pound body composition change and still weigh the same amount.

There is another interesting aspect to muscle. Muscle is denser and takes up much less space than fat. You could therefore maintain the same weight you are now and be 2 sizes smaller by gaining 10 pounds of muscle and losing 10 pounds of fat. Just think, by increasing your muscle, or your lean body mass by 10 pounds, you would increase your metabolic rate by about 350 calories a day. Just gaining 5 pounds of muscle would increase your calorie expenditure by 175 calories a day. That's 63,875 calories or 18 pounds a year.

No wonder as we age it is so easy to gain fat simply by not exercising. Many people are frustrated by the fact that they eat just as they always have, but now they are gaining weight, not realizing that each year without resistance exercise, their metabolisms are slowing down. Exercise is the only way to reverse this effect of aging.

Many people, especially women, are afraid of getting "too muscley," so they avoid strength-training exercises altogether. I know because I used to be one of those people, but that was before I discovered that muscle not only makes you feel better and look more toned, it also greatly increases your body's ability to burn fat. After so many years of starvation dieting, my metabolism needed to be raised from the dead. It seemed as though it was impossible for me to lose weight and keep it off. Replacing lost muscle was crucial to reaching and maintaining my weight-loss goals.

Now, don't worry about looking like the Incredible Hulk. You won't. You'll just look fit because that's what you'll be—FIT! Embark on this strength-training program and watch your body composition start changing. These are great workouts that you can do at the gym or at home. Go for it!

MARTY'S TOP TEN BODY-SCULPTING EXERCISES

From the outset, I want you to know that there are a wide range of exercises both with and without apparatus that you can use to accomplish your goals. I'm well aware of all the fancy machines that are found in health clubs and gyms. And yes, I have seen the infomercials that promise "ripped" abs in 3 weeks. The exercises I have included in this top 10 workout program, however, are those that best target specific muscles in order to give you excellent total body sculpting without spending hours in the gym or buying expensive equipment. All you need are dumbbells. I've done this workout both in gyms around the world and in the privacy of my own home. I truly believe that this program, coupled with aerobic exercise, will give you better results in less time than any program that I have seen.

Depending on your body composition, you could see results in as early as 2 weeks. Be honest with yourself in recognizing that the more body fat there is covering your muscles, the longer it will take to see those muscles resculpting. But be patient! Rest assured that change is occurring, and stay focused on the way your clothes are fitting. You will see a difference!

With that in mind, I am giving you four different methods for using the following top ten exercises to bring about the best results in the least amount of time! And just in case you are like me and are sometimes challenged with time management, one of these programs is called *Crunched for Time? Take 10* that really moves you into high gear. Now that there are no more excuses, let's get started.

One of the following workouts should be done with weights 2-3 days per week, preferably not on consecutive days. For faster results, I recommend 3 days a week with 3 days of aerobic exercise either in between or in conjunction with these routines. As I discuss in chapter 10, even more aerobic exercise is beneficial when you want to burn fat faster. It is also recommended to "cross train," or alternate some of your workouts now and then.

These workouts are designed to work all of the major muscle groups of the body in each workout. Doing these exercises faithfully will produce dramatic results as you reshape your body proportionally. For programs 1-3, allow approximately 30 to 60 minutes for the whole routine. However, before you begin, study the photos carefully and read the descriptions that accompany each exercise to ensure that you are practicing flawless form for each exercise. I recommend 5-10 minutes of walking and stretching to prepare your muscles for weight training.

■ While your intensity levels will depend on which workout you choose, the exercises for all 4 workouts are as follows:

MARTY' S TOP TEN BODY-SCULPTING EXERCISES

TERMINOLOGY OF MOVEMENTS

- A **rep** or repetition is one complete movement of the exercise. Example—1 biceps curl

- A **set** is the number of repetitions repeated without a rest. Example—10 biceps curls

- **Failure** means repeating an exercise until no more can be repeated in sequence without compromising correct form.

1. SIMULTANEOUS STANDING BICEPS CURLS

The most famous muscle on the body is the biceps, which lets you hold a grocery bag in one arm and a baby in the other without dropping either. This exercise develops, shapes, strengthens, and defines the entire biceps muscle, and helps to strengthen the forearm.

POSITION:

Stand with your feet together or a natural width apart with a dumbbell in each hand. Place your arms at your sides and hold the dumbbells with palms facing your body.

MOVEMENT:

Flexing your biceps muscles as you go, and keeping your arms close to your body and your wrists slightly curled upward, curl your arms simultaneously while turning your palms to

face your shoulders until you cannot curl them any farther. Willfully flex your biceps and return to starting position. Feel the stretches in your biceps muscles and repeat the movement until you have completed your set.

FORM:

Don't rock back and forth as you work. Your body should remain stationary—only your arms are moving. Extend your arms without locking your elbows. Breathe naturally.

Workout I	3 sets	8 to 12 reps
Workout II	3 circuits	12 to 20 reps
Workout III	3 to 5 sets	3 to 5 reps
Workout IV	1 set to failure →	→ →

2. MILITARY PRESS

This exercise develops, shapes, strengthens, and defines the entire shoulder muscle group, especially the front area of this muscle.

POSITION:

Stand or sit with feet together in a natural position, holding a dumbbell in each hand at shoulder height, with your palms facing away from your body. Contract your abdominal muscles to stabilize your torso.

MOVEMENT:

Press both arms straight overhead simultaneously until they are fully extended. Keep the weights steady as you slowly lower them to the starting position.

FORM:

Keep your abs tight and your body still as you work. Remember to flex your shoulder muscles on each upward movement, and to feel the stretch on each down position.

Workout I	3 sets	8 to 12 reps
Workout II	3 circuits	12 to 20 reps
Workout III	3 to 5 sets	3 to 5 reps
Workout IV	1 set to failure →	→ →

3. SIDE LATERAL RAISE

This exercise develops, shapes, strengthens, and defines the entire deltoid muscle, especially the medial area.

POSITION:

Stand with your feet hip width apart, knees slightly bent, abdominals tightened, and arms hanging by your sides. Hold a dumbbell in each hand, palms facing in.

MOVEMENT:

Exhale and slowly lift your arms upward and outward, flexing your shoulder muscles as you go until the dumbbells are slightly higher than your shoulders. Pause, inhale, and slowly lower to starting position.

FORM:

Do not swing or use momentum to lift the dumbbells out and up and keep arms slightly bent at the elbows. Use controlled movement, but do keep them moving. Breathe naturally.

Workout I	3 sets	8 to 12 reps
Workout II	3 circuits	12 to 20 reps
Workout III	3 to 5 sets	3 to 5 reps
Workout IV	1 set to failure ➜	➜ ➜

4. BENT OVER ROW

This exercise develops, shapes, strengthens, and defines the upper back (latissimus muscles) and helps to develop the biceps.

POSITION:

Bend over and place your left hand on a chair. Take a dumbbell in your right hand. Make sure your back is straight and parallel to the floor.

MOVEMENT:

Lower the dumbbell so that your right arm is hanging straight down. Slowly but steadily raise the dumbbell to the side of your ribs until your upper arm is parallel or slightly beyond parallel to the floor. Slowly return to starting position. After completing the full number of repetitions, repeat the exercise using the left arm.

FORM:

Contract your abdominal muscles to protect the lower back. Keep your back parallel to the floor. Consciously lower the weight slowly so you are using your back muscles. Keep elbows close to your torso during the contractions. Please note that if you are using heavy weights, you can protect your back even better by placing your knee (same side as working arm) on the edge of your chair or bench for added support.

Workout I	3 sets	8 to 12 reps
Workout II	3 circuits	12 to 20 reps
Workout III	3 to 5 sets	3 to 5 reps
Workout IV	1 set to failure ➜ ➜	➜

5. TRICEPS KICKBACK

This exercise develops, shapes, strengthens, and defines the entire triceps.

POSITION:

Bend over and place your left hand on a chair. Take a dumbbell in your right hand. Make sure your back is straight and parallel to the floor. Bend your right arm at the elbow so that your forearm is nearly parallel to the floor and your elbow is touching your waist.

MOVEMENT:

Keeping your upper arm close to your body as you go, extend your arm as far as possible, and willfully flex your triceps muscle. Return slowly to the start position and feel the stretch in your triceps muscles. Without resting repeat this movement with the opposite arm.

FORM:

Keep your upper arm close to your body throughout the exercise. Don't jerk the dumbbells back. Control your movements. Breathe naturally.

Workout I	3 sets	8 to 12 reps
Workout II	3 circuits	12 to 20 reps
Workout III	3 to 5 sets	3 to 5 reps
Workout IV	1 set to failure ➜	➜ ➜

6. THE LUNGE

This exercise develops, shapes, strengthens, and defines the front and inner thighs, hips, and buttocks. I consider it the best of all leg-shaping exercises.

POSITION:

Stand straight with a dumbbell in each hand. Place your feet a natural, comfortable width apart and look straight ahead.

MOVEMENT:

With head and torso erect and leg muscles tensed, lunge forward with your right leg about two or three feet (depending upon your height), letting your left knee bend so that it stops slightly above the floor. In this position, flex the muscles of your stretched left leg. Make sure that your forward knee never breaks the vertical plane of your forward toe position. As you rise to the start position, flex your quadriceps muscle when you reach the start position. Repeat this movement for the other leg and continue lunging until you have finished the appropriate repetitions.

FORM:

Much of the work of this exercise takes place as you rise from the down position. Your back should remain vertical during the entire movement. In order to get the maximum tension in the working leg, you will have to control your lift-off from your forward foot. Pressing your heel into the ground slightly targets additional intensity to the glute area. This will provide resistance in the working muscle.

Workout I	3 sets	8 to 12 reps
Workout II	3 circuits	12 to 20 reps
Workout III	3 to 5 sets	3 to 5 reps
Workout IV	1 set to failure →	→ →

7. THE SQUAT

This exercise develops, shapes, strengthens, and defines the front thigh (quadriceps) muscle and helps to tighten and tone the buttocks (gluteus maximus).

POSITION:

With a dumbbell in each hand, stand with your feet about shoulder width apart and your toes pointed slightly outward. Let your arms hang naturally at your sides, holding the dumbbells with your hands facing the body. Keep your back straight and your eyes straight ahead.

MOVEMENT:

Feeling the stretch in your front thigh muscles as you begin to bend at the knees, descend to between 45 and 90 degrees. Flexing your quadriceps as you go, return to the start position. Willfully flex your quadriceps and repeat the movement for the appropriate number of repetitions.

FORM:

At first you may not be able to descend to 45 degrees. You may find that you naturally rise on your toes as you descend. If so, you may want to place a board under your heels.

Workout I	3 sets	8 to 10 reps
Workout II	3 circuits	12 to 20 reps
Workout III	3 to 5 sets	3 to 5 reps
Workout IV	1 set to failure →	→ →

8. STANDING STRAIGHT-TOE CALF RAISE

This exercise develops, shapes, strengthens, and defines the entire gastronemius (calf) muscle.

POSITION:

Stand with your feet a natural width apart, your toes flat on the floor, and hold a dumbbell in each hand, palms facing inward.

MOVEMENT:

Flexing your calf muscles as you go, raise yourself as high as possible on your toes. When you reach the highest point, consciously flex your calf muscles and repeat the appropriate number of repetitions for the workout you are following.

FORM:

You have the option of using a thick book or a board under the balls of your feet if you feel that you are not getting a full range of motion. If you think you need even more resistance (especially for Workout I), try doing this exercise by holding the back of a chair for support, and raising and lowering on one leg at a time. (Be prepared for sore calves if you do.)

Workout I	3 sets	8 to 12 reps
Workout II	3 circuits	12 to 20 reps
Workout III	3 to 5 sets	3 to 5 reps
Workout IV	1 set to failure ➔ ➔	➔

9. THE PUSH-UP

If you were to name just one exercise that has the biggest payoff per repetition, the push-up would win hands down (literally) for the upper body. This one exercise develops, shapes, strengthens, and defines the upper and lower pectorals (chest), shoulders, triceps, and much of the body musculature.

POSITION: **VARIATION #1**

Lie facedown on the floor with your hands just outside your shoulders. With your knees on a padded surface, cross your ankles and raise your feet. Keep your back straight and hips locked, making a straight line from your knees to your neck.

MOVEMENT:

Slowly straighten your arms and lift yourself, maintaining a smooth, not jerky, motion. Reverse direction when your arms are fully extended and lower yourself to within 4 to 6 inches from the floor and once again extend your arms smoothly back to the straight arms position. Continue for the appropriate number of repetitions.

Workout I	3 sets	8 to 12 reps
Workout II	3 circuits	12 to 20 reps
Workout III	3 to 5 sets	3 to 5 reps
Workout IV	1 set to failure ➜	➜ ➜

POSITION: **VARIATION #2**

Lie facedown with your hands flat on the floor and parallel with your chest. Keep your head up slightly. Your hand position should be just outside and slightly in front of your shoulders.

MOVEMENT:

Lift yourself by extending your arms in a strong controlled motion so that only your hands and toes are in contact with the floor. This is more difficult than Variation #1.

Workout I	3 sets	8 to 12 reps
Workout II	3 circuits	12 to 20 reps
Workout III	3 to 5 sets	3 to 5 reps
Workout IV	1 set to failure →	→ →

POSITION: **VARIATION #3**

I recommend this variation only for men or very strong women. Putting your feet up on a chair or low table with your hands on the floor places the emphasis on the upper-chest region as well as the front deltoids and triceps. This makes Variation #3 more difficult because you are moving a greater percentage of your body weight.

MOVEMENT:

This is exactly the same movement as Variation #2 except that your toes are balanced on a chair or low table. You may want to position your hands another inch or so in front of your body for better stability.

FORM:

Regardless of which variations you use, be sure to keep your back straight and your legs and abdominals tight so your body forms a rigid line. Don't lock your arms at the top of the movement, and don't allow your midsection to sag or touch the floor at any point.

Workout I	3 sets	8 to 12 reps
Workout II	3 circuits	12 to 20 reps
Workout III	3 to 5 sets	3 to 5 reps
Workout IV	1 set to failure →	→ →

10. THE CRUNCH

This exercise develops, strengthens, and defines the upper and lower abdominal muscles.

POSITION: **VARIATION #1**

Lie flat on your back on the floor. Put your lower legs on a chair or bench or pull your knees up until your legs form an L. You may cross your feet at the ankles. Place your hands behind your head.

MOVEMENT:

Flexing your abdominal muscles as hard as possible, slowly raise your shoulders from the floor in a curling movement until your shoulders are completely off the ground. Do not raise your lower back off the ground. Without losing control or dropping to the ground, lower yourself slowly to the start position and repeat the movement until you have completed the appropriate number of repetitions.

POSITION: **VARIATION #2**

Begin in the same position as Variation #1.

MOVEMENT:

This movement begins the same as Variation #1. Lifting the shoulders off the ground slowly, while tightening your abdominal area, point your left elbow toward the outside of your right knee as in the photo, then return to the start position and repeat the movement to the other side. The twisting motion focuses on the obliques. Continue until you have completed the appropriate number of repetitions to each side.

FORM:

The temptation to lurch off the floor in an effort to gain momentum and make the work easier could cause injury. Lift from the chest area; do not pull on your neck. As you work, visualize your belly button pressing down through your body into the floor. This will help you maintain correct form. Breathe naturally.

Workout I	3 sets	8 to 12 reps
Workout II	3 circuits	12 to 20 reps
Workout III	3 to 5 sets	3 to 5 reps
Workout IV	1 set to failure ➜	➜ ➜

WORKOUT I
BASIC

WHEN YOUR GOAL IS TO TONE, SCULPT, AND RESHAPE YOUR BODY, I personally recommend using a level of resistance that allows you to complete 3 sets of 8 to 12 repetitions with no more than 60 to 90 seconds between sets. If you complete your first set and don't need that long of a rest, take a shorter one. Once you get going, however, you will need a rest. On exercises #4 and #5, you will be doing 10 repetitions with one arm and then changing sides. You therefore won't rest in between side changes.

Refer to the illustrations to guide you through each exercise until your form is safe and controlled. Good form ensures good results. I use this program most often when weight training. I use 10-12-pound dumbbells on exercises 1-7 and 25-pound dumbbells for exercise number 8. If you are just starting an exercise program, it is so important to start at your own level and consistently progress.

1. Simultaneous Standing Biceps Curls	3 sets of 8 to 12 repetitions
2. Military Press	3 sets of 8 to 12 repetitions
3. Side Lateral Raise	3 sets of 8 to 12 repetitions
4. Bent Over Row	3 sets of 8 to 12 repetitions
5. Triceps Kickback	3 sets of 8 to 12 repetitions
6. The Lunge	3 sets of 8 to 12 repetitions
7. The Squat	3 sets of 8 to 12 repetitions
8. Standing Straight-Toe Calf Raise	3 sets of 8 to 12 repetitions
9. The Push-up	3 sets of 8 to 12 repetitions
10. The Crunch	3 sets of 8 to 12 repetitions

WORKOUT II
CIRCUIT

IF YOU WANT A METHOD FOR SLIMMING AND TONING YOUR MUSCLES QUICKLY, this approach is it. It is often referred to as circuit training, in which you do all ten exercises in sequence with **30 to 60 seconds maximum rest between circuits**. If you start with 12 reps, you would do 12 biceps curls, 12 military presses, 12 side lateral raises, and continue without a rest until the end of the circuit. The sequence is then repeated two additional times for a total of three complete circuits. In this program, you perform 12 to 20 repetitions in each set with weights that are lighter than those used in Workout I or III. Once again, if you stick to flawless form, the results are soon to follow.

	Three complete Circuits
1. Simultaneous Standing Biceps Curls	(12-20 reps)
move immediately to #2 without a rest	(12-20 reps)
2. Military Press	(12-20 reps)
3. Side Lateral Raise	(12-20 reps)
4. Bent Over Row	(12-20 reps)
5. Triceps Kickback	(12-20 reps)
6. The Lunge	(12-20 reps)
7. The Squat	(12-20 reps)
8. Standing Straight-Toe Calf Raise	(12-20 reps)
9. The Push-up	(12-20 reps)
10. The Crunch	(12-20 reps)

WORKOUT III
HIGH-INTENSITY WEIGHT TRAINING

IF YOU ARE WEAK OR WISH TO GAIN WEIGHT, I would recommend that you follow the approach of famed Russian strength and conditioning authority, Pavel Tsatsouline. Pavel has recently been employed by our U.S. Armed Forces to create the ultimate strength and conditioning programs for the men and women of the United States Marines. This workout is particularly good for men desiring to dramatically increase muscle size and strength, and for women desiring to greatly increase strength. Training at a high weight intensity with fewer repetitions does encourage muscle growth, but not as much in women as in men. So when your goal is strength and/or size, Pavel recommends doing *3 to 5 sets* of *3 to 5 repetitions* with as much resistance as you can safely control while maintaining *flawless* form. This is extremely difficult and requires rest periods of up to 2 minutes or more between sets, making it very time consuming, but yes, absolutely, *it does deliver results.*

1. Simultaneous Standing Biceps Curls	3 to 5 sets of 3 to 5 repetitions
with 2 minute rest between sets	
2. Military Press	3 to 5 sets of 3 to 5 repetitions
3. Side Lateral Raise	3 to 5 sets of 3 to 5 repetitions
4. Bent Over Row	3 to 5 sets of 3 to 5 repetitions
5. Triceps Kickback	3 to 5 sets of 3 to 5 repetitions
6. The Lunge	3 to 5 sets of 3 to 5 repetitions
7. The Squat	3 to 5 sets of 3 to 5 repetitions
8. Standing Straight-Toe Calf Raise	3 to 5 sets of 3 to 5 repetitions
9. The Push-up	same as other routines
10. The Crunch	same as other routines

Example: If you normally use 8- to 10-pound dumbbells in Workout I, you would try 15-20-pound dumbbells, determined by your ability to safely control each repetition. If you normally use 30-pound dumbbells in Workout I, you would try 45-50-pound dumbbells for this workout again, with controlled form.

WORKOUT IV
TIMESAVING

CRUNCHED FOR TIME? TAKE 10!

THE NUMBER ONE EXCUSE GIVEN by non-exercisers is: "I just don't have time." Remove all the typical distractions (talking, daydreaming, watching TV), and the time many people spend weight training usually boils down to less than 20 minutes per session. To help you take advantage of those time-crunching days, this workout is designed so that you can do it in a small amount of time but still get big results. There's even an added benefit to occasionally incorporating this workout into your regular schedule. You see, muscles adapt to any routine quickly, and results can be intensified by varying your exercises, intensities, or weights. By omitting your rests in between sets and using lighter weights, the following 10-minute dumbbell routine does just that.

Do one set of each of the 10 body-sculpting exercises for as many repetitions as you can in flawless form, then immediately move on to the next exercise without resting. Once you finish the final exercise, repeat the entire sequence again until you reach your 10-minute goal. Stick the outline where it's handy, and the next time priorities cut your exercise time to shreds, you will already have your back-up plan.

TO START:

- Choose your dumbbells 50 to 70 percent lighter than if you were doing your standard workout: Workout I. Do not worry about your weights being too light. Lifting lighter weights without resting between sets will feel as challenging to your muscles as lifting heavier weights with the longer rest periods. Because the weights are lighter, you must remind yourself to concentrate on using safe and proper form. Do the movements quickly, yet controlled.

Without the rest time in between sets, you should find yourself working aerobically, or with oxygen. However, as you may be a beginner to weight training or are simply increasing your level of difficulty, you could find yourself becoming anaerobic, or out of oxygen. (If you run out of air, by all means, rest a minute to regain your oxygen.) Never continue to exercise if you feel dizzy or nauseous. You must learn to distinguish the difference between challenging your fitness level and pushing too hard.

Remember, you only have 10 minutes, so keep it moving.

Exercise			
1. Simultaneous Standing Biceps Curls	to failure ➜	➜	➜
2. Military Press	to failure ➜	➜	➜
3. Side Lateral Raise	to failure ➜	➜	➜
4. Bent Over Row	to failure ➜	➜	➜
5. Triceps Kickback	to failure ➜	➜	➜
6. The Lunge	to failure ➜	➜	➜
7. The Squat	to failure ➜	➜	➜
8. Standing Straight-Toe Calf Raise	to failure ➜	➜	➜
9. The Push-up	to failure ➜	➜	➜
10. The Crunch	to failure ➜	➜	➜
	(start sequence again)		

WORKOUT V

BONUS AB-BLASTING WORKOUT!

WHEN GREAT-LOOKING ABS are important to you, here is an ab-blasting workout that will blast those abdominal-gizmo infomercials right out of the water! Though it appears that infomercials promising washboard abs in as little as 3 weeks may be replacing baseball as America's number one spectator sport, I'm confident that this workout can help you achieve the kind of muscle strength, tone, and definition you want without wearing out your pocketbook. I personally counted on these moves to help me strengthen stretched muscles to regain my pre-pregnancy shape. Here are 10 great exercises for those of you really serious about your midsection.

THE EXERCISES

Physical fitness should enhance the quality of our lives and prevent us from injuring ourselves. Many sit-ups and certain variations of leg raises can place great stress on the lower back and may lead to injury. I have intentionally avoided such exercises and focused on the one I believe to be most effective—the "crunch." Unlike many midsection exercises, the crunch works the abdominal muscles without great involvement from either the hip flexor or leg muscles and places very little stress on the lower back. This makes it the ideal exercise to sculpt and develop the abdominals directly. The success of the abdominal crunch, however, is solely determined by the technique that you use. In fact, by following proper technique, you'll soon discover that it's not the number of crunches you do; it's how you do each crunch that matters. So please pay special attention. These are the hard and fast rules on the proper way to crunch to give you the results you've been looking for.

RULE #1: **Stay low.** Rise up only as far as you need to. After you raise your torso beyond 45 degrees off the floor, you stop emphasizing the abdominal muscles and begin using your back and leg muscles to pull yourself up. That's not doing your abs any good. To save time, help prevent injury, and keep continuous tension on the muscles, curl up only until your shoulders clear the floor by a few inches.

RULE #2: **Squeeze up**. Because of misinformation, many people think the premise of the crunch is to curl the shoulders up and forward. This is improper technique and causes people to tighten muscles of their upper back and neck, placing unnecessary strain on these areas as they train. Instead, contract your stomach muscles by drawing your ribs to your hips, as if your midsection were an accordion. Your shoulders will still lift off the floor. Only now they will simply be going along for the ride.

RULE #3: **Suck it in**. Drawing in your stomach isn't just a way to appear to have a flatter stomach. It can actually help you get one faster. Holding in your abdominal muscles as you go helps target the muscles to make the exercise more effective.

RULE #4: **Defy gravity**. Consciously lower yourself to the floor. Letting yourself fall back down during the lowering phase of a crunch puts your abs through only half a workout. By slowly lowering yourself to the floor, you will get twice the results from the same exercise.

With these rules in mind, here are ten great exercises that work all three muscle groups that make up the midsection (the upper abs, lower abs, and obliques). Do one set of each exercise, waiting only 30 seconds between each set. (To protect your lower back, always be sure to perform each exercise on a mat or padded surface.)

EXERCISE #1: THE CRUNCH

Lie on a carpeted floor or mat and place your hands gently behind your head. Then lift your knees up and place your feet flat on the floor about a foot from your hips, bringing your knees together and turning your toes in. Start the exercise by pushing your lower back down—into the floor—and begin to roll your shoulders up for a count of "one thousand one, one thousand two." While flexing your abs as hard as possible, your hips and knees will remain stationary. Your shoulders will come off the floor only a few inches. After you've achieved peak contraction, slowly lower your shoulders back down to the floor. This is one complete repetition. Initially you will repeat this exercise for 1 set of 5 high-tension repetitions. Gradually, over the course of the next 5 weeks, you will work up to 10 repetitions per set.

If you follow and achieve a peak contraction on each repetition, you will never need to go beyond 10 repetitions in any given set. This is also true in the 9 variations that follow, in which only the foot and leg positions change. These changes cause different areas of the abdominal muscles to be emphasized, thus ensuring you a complete workout for every part of the abdominal musculature.

CRUNCH: **VARIATION #2**

In variation #2 the lower abdominal and oblique muscles on the sides of the waist are affected by opening the knees wide and pressing the bottom of your feet together. As before, raise and lower, holding to peak contraction for a count of "one thousand one, one thousand two."

CRUNCH: **VARIATION #3**

Raise your legs with your knees together and toes turned in to work the lower abdominal rows. Follow the same procedure as above, achieving peak contraction for a count of "one thousand one, one thousand two." (You didn't know 5 reps could be this intense, did you?)

CRUNCH: **VARIATION #4**

Variation #4 exercises obliques and lower abs from a new angle. Legs raised, knees opened wide, and place the bottom of your feet together. Once again, raise and hold peak contraction, then lower. (5 to 10 repetitions.)

CRUNCH: **VARIATION #5**

The center of the midsection is worked when the legs are straight up in the air. As in previous sets, slowly raise, contract (flex), and lower.

CRUNCH: **VARIATION #6**

Lock the knees and spread the legs to work the upper abdominal row. Think about each repetition. Concentrate on the muscles being worked.

CRUNCH: **VARIATION #7**

The left leg is held straight, six inches off the ground, while the right leg is bent 90 degrees. (A very difficult variation, but the results will be worth it.)

CRUNCH: **VARIATION #8**

Same as variation #7, only the leg positions are reversed.

CRUNCH: **VARIATION #9**

The obliques are worked by bending your knees and raising and lowering the upper body in a controlled turn at the waist. A partner may be used to help keep your knees down.

CRUNCH: **VARIATION #10**

Identical to variation #9, only reverse position and raise and lower the other side of your body.

From improving how we feel, how we look, and how we think to increasing our life expectancy, the ability to exercise and literally change the quality of our lives is something we should be thankful for. Exercise is not only an anti-depressant, but it increases the quality of our sleep. Another one of its greatest benefits is that it's never too late to start! Even beginning an exercise program in our later years can greatly improve our health and mobility. It is vital that we continue to use our bodies so that we don't lose our physical capabilities. We should plan on growing old well able to enjoy an active lifestyle!

IT'S A LITTLE KNOWN FACT...

Not only does weight training reduce your blood pressure, lowering your risk of cardiovascular disease, but a landmark study performed by researchers at the USDA Human Nutrition Research Center at Tuft's University revealed that weight training has a tremendous effect on aging. In only 8 weeks, 10 frail men and women between the ages of 86-90, all of whom had chronic diseases or disabilities, increased the strength of their leg muscles by a dramatic 174 percent without any injuries! They also increased their balance and walking speed.[6]

According to the American College of Sports Medicine, weight or resistance training is the only type of exercise that can substantially slow and even reverse the declines in muscle mass, bone density, and strength that were once considered inevitable consequences of aging.[7]

As always, consult your physician before beginning this or any exercise program.

10 FAMILY FITNESS TIPS

CHAPTER

8

Marty Copeland's

HIGHER FITNESS

EIGHT

10 FAMILY FITNESS TIPS

LET'S FACE IT. There is nothing that can take the place of quality time with your family. Why not incorporate family time with fitness time and make the whole family happier and healthier? Maybe you'll want to start with just once a week and make 2 or 3 times a week your goal. The most important thing is that you start! Even if it's just 1 night a week, begin scheduling time into your weekly schedule for family fitness. From hand washing the car to walking the dog, there are many ways to get the family moving. Even a night of bowling will burn a lot more calories than a night of television (assuming you go after dinner and avoid the colas and junk food). Eating less junk will also save you a lot of money.

The most important part of family fitness is that everyone enjoys it. Make every effort to be fun and flexible to what the kids want to do. This is a great opportunity to build your kids' self-esteem and brag on their strengths. If their athletic skills are wanting, brag on their people skills or their intelligence in the game. Make family fitness time off limits for quarrels or confrontations and encourage fun topics. Be creative. Once you set a time, mark it on a calendar where everyone will see it, and then talk it up! It won't take long before your entire family will look forward to this time together. Here are some fun fitness ideas to get your family fitness time off to a great start!

1. **Find a nearby park and play.** Bike, walk, or drive to a nearby park and take a Frisbee, football, or soccer ball and play. You'll not only encourage exercise, but you'll also make some great family memories. If you go on Saturday or during the summer, try taking a healthy picnic lunch on occasion. Make sure helmets or other necessary safety gear is worn and review all relevant safety rules. Always make safety a priority and don't forget the water.
2. **Walk around the neighborhood.** Walking can be great exercise, and all you need are some good walking shoes. Push the kids in a stroller or let the dog lead the way. Better yet, plot a strategic course and time yourselves. Buy a stopwatch and let the kids take turns timing you. You might even have a nearby park or jogging path. If safety is a challenge, drive to a track or walk at the mall. Wherever you go, start at your own fitness level and be progressive. Make it a little more challenging each week by increasing either your distance or your speed. Don't do too much too soon!
3. **Go military.** Take turns being the captain in the army and bark orders at one another. Make one another do a realistic amount of calisthenics. You can run in place to the count of 30. You can do sets of jumping jacks, sit-ups, or push-ups. If necessary, remind the kids to take it easy on Mom and Dad and make sure not to

overdo it. See if you can keep a tally of everyone's personal record. If you did 20 sit-ups on your last military night, see if you can do more. Let the girls do push-ups on their knees, since they normally have less upper body strength. Try to make things fair and not too competitive.

4. **"Get jiggy with it."** Play some fun, kid-friendly music and start dancing. Music can get you moving and is a great way to burn some calories! If you're too embarrassed to dance, I encourage you to get over it! My family has so much fun laughing at me trying to do this hip-hop step that I simply cannot do! They think it's hysterical. Their laughter is much more valuable than silly pride.

5. **Jump rope together.** Now this can be an advanced workout and should be done very carefully. If you are out of shape, overweight, have weak joints, or are simply a beginner, you may want to walk in place or gently bounce from foot to foot instead of jumping. If you have your doctor's permission to do such an activity, jumping rope is a great workout, and your kids might get a kick out of seeing Mom and Dad jumping rope. Even if you don't own a jump rope, you can pretend you are jumping. The best way to do this workout is to count repetitions. If you run out of air, walk around and then start again. You'll want to stretch your calves really well before you start. The

first time I jumped rope for very long, my calves were screaming at me the next morning. Take it slow!

6. **Invest in a basketball hoop.** Basketball is great exercise! The serious exercisers can play one on one, and the beginners can learn to dribble. Buy a book that teaches shooting games such as "HORSE."
7. **Tend a garden together.** Gardening can be a great workout: pulling weeds, digging, hauling around a watering can, etc. It's not only a great way to get exercise, it's also great fun to watch a garden grow.
8. **What happened to chores?** Give each one of the kids some chores that will get them out from in front of the TV. Rotate their chores to be fair. If your daughter prefers sweeping off the driveway and your son prefers washing the car, let them swap chores. If everyone does all their chores for a couple of weeks, reward them. Tell them how proud you are of their diligence and take them for "kids' night out." Go play video games, go ice-skating, play putt-putt golf, or rent a special movie and let them have a few friends over. A little work will teach them responsibility, and doing chores together will make them feel like a team.
9. **Just gotta watch TV?** As you may have guessed, I think TV can be an enemy to an active lifestyle. Ironically, however, one of my favorite things to do is to get into my PJs by about 8:00 P.M. and watch a good movie. Vegging out in front of the TV can be completely relaxing. Of course, the kids bouncing

across the bed does affect the mood somewhat. So how can you use TV time to get more fit? Pick a favorite TV show that comes on before you get too tired and compete against the clock. Who can do the most sit-ups, push-ups, leg lifts, or lunge walks during the 2-minute commercial break? Use the next break to drink an extra glass of water. (The loser gets the water, of course.) Alternate exercises during each break. Let the winner get out of a weekly chore.

10. **Relax together.** This too can be done in front of the TV but would be much better with some relaxing music. Whether you prefer worship music, instrumental, or classical, put on something that will relax you without putting you to sleep. Let each family member pick some favorite stretches that are both safe and effective.

- Do some gentle neck rolls. Starting on one side, hold in a gentle stretch, then gently roll your head forward and to the other side and hold the stretch again. Repeat 4 or 5 times. Do not roll your head to the back.

- Stretch your back. Get on your hands and knees. Arch your back up toward the ceiling like a stretching cat. Hold for 5 seconds and release until your back is straight. Repeat 5 times. Do not arch your back down toward the floor.

- Stretch your hamstrings. Sit up straight on the floor and straighten your legs in front of you. Raise your hands toward the ceiling and with a straight back gently bend forward at the waist. Do not bounce. This can be both painful and somewhat comical if your muscles are very tight. This stretches your hamstrings (the back of the legs) and the back. (When I haven't done this in a while, it feels as though the back of my thighs are nauseated.) With consistent practice your flexibility should improve. Improved flexibility will make you feel better. Take it slow and add some new stretches to your workout.

IT'S A LITTLE KNOWN FACT...

Turning off the tube can help you lose weight. A recent study shows that people who watch two extra hours of TV a day have a higher body fat percentage than those who don't, even when both groups exercised the same amount. Making an effort to be more active during your leisure time will affect your waistline.[8]

ON-THE-PHONE FITNESS

9

Marty Copeland's

HIGHER FITNESS

NINE

ON-THE-PHONE FITNESS

THE AVERAGE AMERICAN SPENDS over three hours a week talking on the telephone (outside of work). Three extra hours a week dedicated to exercising would do wonders to our fitness level, but we can't just hang up on our friends and relatives and hit the pavement. Whether we just have to miss our workout or we just want to do a little extra, here are a few examples of how we can tap into our phone time at home or at the office to literally change our bottom line.

1. DEEP BREATHING

Lowers your heart rate and relaxes your entire body. Inhale deeply and slowly, extending your abdomen. Exhale slowly. Repeat until you feel relaxed. Be sure you are holding the microphone end of your telephone receiver away from your mouth. (You wouldn't want to give the cable guy the wrong impression.)

2. NECK STRETCHES

I recommend doing these while on speaker-phone or pay special attention to your receiver position. Gently stretch your head to the left side. To better isolate the neck muscles, keep your chin tucked under. Slowly roll your head forward and hold, then rotate to the other side. Hold each stretch about 10 seconds. Repeat 4 or 5 times. Do not roll your head to the back.

3. SHOULDER SHRUGS

Sitting with your back straight, arms down to your side, lift your shoulders up toward the ceiling. Squeeze firmly and hold for the count of 10. Gently lower your shoulders down until you feel a nice, easy stretch. Repeat as desired. Personally, most stress and tension tries to store itself in my shoulders. This is one of my favorite exercises.

4. DESK PUSHERS

Whether sitting at a table or desk, these will help strengthen and tone your bicep area. Make sure your back is straight and your lower back is touching the back of your chair. Put both feet on the floor, placing your hands, palms up, underneath the table in front of you. Your arms should be bent at a 90-degree angle with your elbows next to your side. Press firmly and up, getting a nice contraction in the bicep muscle. Hold until muscle tires. If you are very strong or your desk is very weak, beware not to send your desk-top flying.

5. FANNY FIRMERS

Sit upright in a chair with your back well supported. Tighten the buttocks muscles by squeezing very firmly. Hello glutes! Hold for a count of 10 and then relax. Repeat 10-20 times. Increase your repetitions as you get stronger. You can do these in the car or any time your bottom gets tired of sitting.

6. STANDING CALF RAISES

Remove your shoes for the full affect of stretching. Stand with your feet shoulder-width apart, toes facing forward. Hold on to your desk or table for support. Keep your back straight and slowly lift up onto your toes (as high as you can) and hold for a count of 3, then slowly lower back down. You can also perform these with your toes slightly pointing in and then slightly pointing out. Altering your foot position will help work different parts of the calf. It is important to keep the angle small. Start with a few of these and work your way up to 20 or 30 repetitions. If you are used to doing these with very heavy weights, you can try standing on one foot for 10–20 repetitions, then switching legs.

7. PLIES

Yes, this is great for guys, too. Stand with your feet a little wider than your shoulders, toes pointed out at a 45-degree angle. Slightly bend your knees and keep the hips in a neutral position as you stand upright. Avoid arching your lower back. Let your knees point the same direction as your toes (Do not let them extend past your toes.) Continue lowering your body. Do not let your thighs go lower than parallel to the floor. Return to starting position. This should be a slow and controlled movement with your torso moving up and down in the center of the movement. I do understand that these may be uncomfortable and inappropriate in a crowded office.

8. SIDE LEG LIFTS

Stand on one leg with the knee slightly bent (hold on to your chair or desk if you need help balancing). With the feet facing forward and back straight, slowly lift the opposite leg and bend at a 90-degree angle. Lift leg *directly* to the side, hold a few seconds, then lower. Repeat until muscle fatigues. You can add resistance by placing your hand on your outer thigh and pressing against it as you lift. Don't forget to do both sides!

9. HAMSTRING STRETCHES

Stand with your feet shoulder-width apart, toes pointing forward and knees slightly bent. Place your hands on your desk and gently bend forward at the hips until you feel a mild tension in the back of your upper legs. Hold the stretch for at least 10 seconds. Do not bounce, but hold the stretch static. Always keep your back straight.

10. LUNGES

Turn sideways and rest your hand on your desk or a stationary chair back for support. Stand with your feet shoulder-width apart, feet facing directly forward. Take a big step forward with one foot and lower your body toward the ground as you bend your front knee to a 90-degree angle. Keep your torso upright and in the middle of your stance. Do not let your front knee go over your toes. Return to the starting position by pressing your front foot off the floor (pressing through the heel). Alternate legs. Start with just a few of these and increase as your strength increases. Only lower you body as far as your fitness level will allow. Warning: This really works those glute muscles and can make you very sore.

IT'S A LITTLE KNOWN FACT...

How can little changes make a big difference in creating a healthier lifestyle? Think about this the next time you are toning those "glutes" while doing your on-the-phone regime. Either by eliminating 100 calories from your diet a day, or by burning 100 extra calories a day, you could lose 10 pounds of fat in one year. If you modified both your diet and activity levels, and you were consistent, that would mean a 20-pound fat loss in one year. Just by skipping that Coke you have every day at the office, you would save at least 650 calories in five days. At 50 calories per tablespoon, just switching from whole cream to a nonfat, lower-calorie version in your coffee can help save your waistline. Taking the stairs instead of the elevator may not seem like much now, but over the period of a year, it can make a huge difference. Even standing and moving around while talking on the phone burns more calories than sitting. All of these small changes create a healthier and more active you!

CHECK POINTING YOUR PROGRESS

10

Marty Copeland's
HIGHER
FITNESS

TEN

CHECK POINTING YOUR PROGRESS

WE OFTEN CATEGORIZE PHYSICAL ACTIVITY as very separate from spiritual activity, yet God instructs us to take care of our bodies. God made our bodies to function a certain way, and healthy eating and exercise habits are an important key to maximizing our potential. It is crucial to our health and fitness goals that we continue to examine the relationship between our spiritual, emotional, and physical development. Certainly, our spiritual development affects our emotional stability, as does our emotional stability affect our daily decisions. And, of course, it is our daily decisions that determine our lifestyle. An increasing awareness of all three aspects is required for continued improvement.

Once we establish a foundation of spiritual growth and emotional stability, it is much easier to effectively implement the laws that govern our body's ability to lose weight and get fit. If progress slows, stops, or regresses, we should reexamine our respect for those laws. Without even a hint at being all-inclusive, this is an excellent checkpoint list that will help you stay on your road to success.

Are you hungry for God?

If we are not cultivating a personal relationship with God, there will be dissatisfaction in our lives that can manifest itself in numerous ways. From feeling frustrated to feeling unloved, from feeling tired to feeling hungry, a lack of spiritual fulfillment can directly affect our relationship with food and our ability to exercise. This is one of the main reasons Americans spend over thirty billion dollars a year on weight-loss products and continue to get fatter. Food cannot satisfy spiritual and emotional hunger.

God's presence or character is described for us in Galatians 5:22-23 as love, joy, peace, patience, kindness, goodness, faithfulness, gentleness, and self-control. It is God's desire to fill us. In the book of Ephesians, He says to "flood us" with His love, His character, Himself. He does this as we spend time in His presence. As we "fill up" with God, our character begins to take on His character. Is spending time with God and developing His character worthwhile? This very point is crucial to our success. There is no temptation that exists that one of these characteristics will not overcome. A fulfilling relationship with God, balanced by a healthy relationship with others, is vital to our total well-being. It is God who made us that way.

CHECKPOINT 2

Are you hungry for healthy relationships?

Whether we are surrounded by dysfunctional people or find ourselves alone, we hunger for healthy relationships. What is sometimes termed "emotional hunger" is often a result of a lack of relational fulfillment. There may be people in our lives whom we experience many problems with, and we therefore are lacking in fulfilling relationships. Sometimes we ourselves are so dysfunctional that "healthy" people may be avoiding us, or we them. Loneliness can be very painful. Whatever the cause for this void in our lives, the buck stops here, and we must first examine ourselves. Where do we start?

Our first step is always to refer back to Checkpoint #1, because it is there that we get our strength and guidance from God. Then, we, with God's help, of course, walk in His love. We can and should pray for our relationships, but not just that others will change or do what we want them to do. Love is not selfish. We should pray that God bless them and that His will be done in their lives. It is by instruction of God Himself that we are to focus on changing ourselves. How does this affect our emotional health? The power of love is twofold. First, it constantly rewards us with the joy of giving. Second, it gives back to us. The better we treat others, the better they treat us. No one can resist the nature or character of God. As we continue to spend more time with Him, we continue to spend more time with love.

1 Corinthians 13:4-8 in the *Amplified Bible* tells us about love. How would it affect our relationships if these words were describing us? To emphasize this point, I will put my name in the text instead of "love." Put your name in there as you read and evaluate your attitude toward others and their attitude toward you.

Love (Marty) endures long and is patient and kind; (Marty) never is envious nor boils over with jealousy, is not boastful or vainglorious, does not display (herself) haughtily.

(Marty) is not conceited (arrogant and inflated with pride); (Marty) is not rude (unmannerly) and does not act unbecomingly. (Marty) does not insist on (her) own rights or her own way, for (Marty) is not self-seeking; (Marty) is not touchy or fretful or resentful; (Marty) takes no account of the evil done to (her) [(Marty) pays no attention to a suffered wrong.]

(Marty) does not rejoice at injustice and unrighteousness, but rejoices when right and truth prevail.

(Marty) bears up under anything and everything that comes, (Marty) is ever ready to believe the best of every person, (Marty's) hopes are fadeless under all circumstances, and (Marty) endures everything [without weakening].

(Marty) never fails.

Wow! Can you see the power in that? How would family, friends, and co-workers respond to you if you acted like that? This is not a pretense of trying to be someone that we are not. No! God despises hypocrisy. This is an act of yielding to the power of God's love on the inside of us. It is God's love in us that enables us to act loving. As we spend time with God, we should become progressively more like Him, and He is love.

God wants to love us, to counsel us, to heal us, and to bless us so that we can enjoy a wonderful life and be a blessing to others. We don't bless others by judging or condemning them. We bless them by loving them. Simply learning to love others helps us to get our minds off ourselves. Many problems are instantly solved the moment we begin getting our minds off of ourselves. Many emotional challenges stem from our response to the words and actions of others. It is developing our capacity to love with the God kind of love that strengthens us and empowers us to stabilize our emotions. Love is a choice. Love forgives and lets it go. Love is a powerful force.

The proof is in the proving.

God's will, His desire for us, is already in existence. The book of Romans, chapter 12, verses 1-2, tells us, and I summarize, that presenting our bodies to God and renewing our mind to His promises is how we "prove His good and acceptable and perfect will for our lives." For example, we see clearly in the Scriptures that it is His desire, His will, for us "to prosper and be in health, even as our soul prospers." There are many scriptures expressing God's desire for our health and prosperity, and He wants us to establish their truth by experiencing them in our everyday lives.

Presenting our body to God includes submitting our eating and exercise habits to Him. This is where I had missed it for so many years. Yes, I wanted to be free from the bondage of obsessing about my weight, but I didn't know God's will for me in losing weight. Not only was I not consistently seeking God's wisdom and submitting my diet to Him, but I was trying every fad diet known to man, literally abusing my body. And then there was my motive for weight loss. My primary focus in life was becoming a certain size. There is no question that a strong healthy body looks good, and there is nothing wrong with being a certain size, but God's will for our lives is excellent health, not vanity.

While God is always ready and willing to act on our behalf, our goals must line up with His in order to access His spiritual blessing. It is what we decide to do with our minds and bodies that determines whether or not we activate His power.

CHECKPOINT 4

Should you change your mind?

One major aspect that will determine our success at any endeavor is our ability to change how we think. It is how we think about ourselves that has gotten us where we are today. It is how we learn to think about ourselves that will get us where we want to be tomorrow. In the above scripture, verse 2 says that it is a "renewed mind" that causes transformation. A renewed mind makes the difference between just gaining knowledge and successfully applying it to our lives.

I challenge you to do the following assignment to prove the truthfulness of this scripture and its ability to change your thought life. Refer to Checkpoint #2 above and read 1 Corinthians chapter 13, verses 4-8, out loud with your name in it as we did before. Only this time, read it 10 times in the morning and 10 times right before you go to bed. I know you might not want to do that, but make yourself. Do it every day for 14 days and evaluate for yourself the effect it has on your thinking, words, and actions. Two weeks is not long, but it is long enough to confirm that there is transforming power in filling your mind with God's words. First, you will begin to think differently about your potential. Then, you will begin to see yourself differently. Pretty soon, you will begin to act differently. As you act differently, over time, people will begin to respond to you differently. It is amazing how much more effective it is to change ourselves than to try to change those around us. This is a very valuable weapon in defeating challenges that you face. You can, therefore, change your thinking and conquer an unhealthy lifestyle through the power of God's Word. It is God's will for you to live a long, healthy life!

Is your heart troubled?

In Luke 21:25, Jesus is talking about the signs of the end of this age. He talks about the distress of nations and the perplexity and bewilderment of men because of the things happening on the earth. One can only be reminded of the fear and perplexity of so many, resulting from the acts of terrorism beginning on September 11, 2001. In verse 26, He says that there will be men whose hearts will fail because of fear. Research has shown that a fearful mind can caused a stressed-out body, and a stressed-out body can cause heart problems. There is also the very relevant point that when our minds are dominated by fears that affect our decisions, we are sometimes prevented from making the lifestyle choices that we know are good for us. An example of this would be an emotional eater who wants to eat right but turns to food when stressed or lonely. Again, the health of our heart is compromised. Worry, poor eating habits, and poor exercise habits are all contributors to heart problems. It is no coincidence that heart disease is the #1 killer in America today.

It is quite a blessing that we can dramatically affect the condition of our heart through healthy eating and exercise choices. Concerning fear and worry, however, what are we to do? In John chapter 14, verse 1 (KJV), Jesus says, "Let not your heart be troubled: ye believe in God, believe also in me." In John 14:27

(KJV), "Let not your heart be troubled, neither let it be afraid." It would be ridiculous for Him to command us to do something that we are unable to do. Not letting our heart be troubled is clearly a conscious decision that we must make. Philippians 4:6-7 in the *Amplified Bible* says, "Do not fret or have any anxiety about anything, but in every circumstance and in everything, by prayer and petition (definite requests), with thanksgiving, continue to make your wants known unto God." Certainly challenges are inevitable, but look what happens when we choose not to worry and instead ask God to help and thank Him for it.

Verse 7 says, "And God's peace [shall be yours, that tranquil state of a soul assured of its salvation through Christ, and so fearing nothing from God and being content with its earthly lot of whatever sort that is, that peace] which transcends all understanding shall garrison and mount guard over your hearts and minds in Christ Jesus." Evidence of a heart that is trusting God is a peace that passes all understanding. It is this peace that literally guards our heart and mind. Living a life filled with peace instead of fear and worry can have an enormous impact on both our health and the quality of our life. Make the lifestyle decision to not let your heart be troubled!

CHECKPOINT 6 Be accountable and socialize.

As we discussed in Checkpoint #2, healthy relationships are very important. Studies suggest that social interaction plays a significant role in the amount of exercise we get. If we can combine physical activities with friends and fun, we are much more likely both to exercise more and to enjoy it more. This is assuming, of course, that you don't substitute exercise for conversation. If we are married or have children and need our leisure time to be family time, we can pull many ideas from chapter 8, "Family Fitness Tips." Whether we take up tennis and play with friends, get a family membership at the YMCA, or join the nearest fitness club, there are many ways to combine at least one workout a week with social interaction. The added enjoyment and accountability we derive from such a situation can be invaluable.

For several years now, I have adapted my workout schedule to fit my lifestyle. When I first got married, I worked part time, belonged to a gym, played some occasional tennis, and played on a softball team. I found that the social interaction made the exercise more enjoyable. After I had my first child, well, you know, my priorities changed. I worked out at home with workout videos while my daughter slept. Sometimes in the evenings I would jog. After a year or so, I joined the YMCA to take some aerobic classes just to get out of the house. I enjoyed jogging by myself to have some quiet time to think, but I missed some of the social interaction. I didn't experience a sense of accountability at the Y, because I never really took the time to get to know anyone very well. I was always rushing home to get back to my daughter.

My schedule became more flexible when my daughter started school. A small workout facility opened where my husband and I worked. It was very close to where we live, and it was free. My sister-in-law, a friend of hers, and I started working out together in the mornings. Of all the schedules that I have ever had, this one seemed to work the best. I realized that working out in the mornings kept me from "blowing it off" later in the day. After several months, my sister-in-law changed her schedule, but my friend and I are still workout buddies. Though I occasionally disappeared for a few months, while pregnant or nursing, I would always return, knowing that my faithful partner would be there. I daresay that I have grown greatly over the years in the area of discipline. Nevertheless, there have been countless busy days when the only thing getting my gluteus maximus to the gym was knowing someone there was expecting me. The point is, while exercise is a personal commitment of hard work, creating an environment of accountability can make it easier, and social interaction can make it more enjoyable.

We even amazed ourselves one summer when we decided that if we worked out really early in the morning, our husbands could stay home with the kids, and we would be back before breakfast. I will never forget the feeling of finishing a hard workout, grabbing my water bottle, and walking outside to see the sun rising. I would be home before the kids got up. I felt invincible. I felt like that Army commercial, "We do more by 6:00 A.M. than most people do all day. Be all that you can be, in the AAAAArmy. . . ." Okay, Okay, I digress. My workout partner and I have proven over time that we are committed to our goals no matter what. Together or alone, we are committed to exercise!

Don't be discouraged if your lifestyle or pocketbook dictates that you go it alone. I have spent countless hours "going it alone" myself. You can rest assured that there are thousands out there who are doing it, too. Many people prefer it that way, and there is a very strong sense of accomplishment from achieving your fitness goals by yourself. If you want more social interaction and your lifestyle allows it, go for it. Be it a workout partner, personal trainer, or aerobics class, there are great benefits to be gained from accountability and social interaction. The most important thing is to be flexible to what you can do right now.

When should you exercise?

Many people have asked me, "When is the best time to exercise?" My answer is, "Whenever you can." Research shows that there are many benefits to exercising in the morning. Exercise revs up your metabolism and can help keep you burning more calories for hours. I know for me that exercising in the morning works best for many reasons. First, it ensures that the demands of my day won't cause me to cancel it. Second, it has a tremendous effect on how I feel. Not only do I feel more energetic, but I also have a sense of accomplishment that positively affects my mood. I experience better concentration and focus throughout the day, and can even tell a difference in my appetite. Not only do I experience a decreased desire for certain foods, but I am more motivated to make better food choices because I am more aware of not sabotaging my own fitness efforts.

Being able to accomplish all of these benefits makes morning workouts a worthwhile goal. An added benefit of regular exercise is that the quality of our sleep improves, and we sleep sounder. This means that getting up an hour earlier to exercise does not necessarily mean that we have to go to bed an hour earlier, because as the quality of our sleep improves, we may require less sleep. But really, what would it hurt to sacrifice a little late-night TV for a stronger, healthier, and more rested body?

Now, I am going to turn right around and tell you that if you just can't work out in the morning, fine, then don't. I know people who work out for 30 minutes every day during their lunch hour at work. I know people who worked out during their baby's afternoon nap time. I know people who work out in the evenings. One of my good friends works out whenever she can, be it mornings or evenings. Her goal is simply to get in 4 workouts per week. She is disciplined enough not to need a schedule. Remember, *when* you work out is not nearly as important as *if* you work out.

CHECKPOINT 8

How much is enough?

While 3 workouts a week is both a great place to start and great for *maintaining* an intermediate to advanced fitness level, there seems to be something powerful about each additional workout for losing weight and changing your body. Getting an unfit body fit quite honestly requires some hard work. My personal advice to people desiring to lose weight or reshape their body is to work out aerobically a minimum of 4 days a week including weight training on at least 2 of those days. For some, I go on to recommend that they just go ahead and work out aerobically 5 times, Monday through Friday, for even faster results.

Really, I have found that for many people, forming the habit of working out during the workweek on consecutive days is easier than breaking on Wednesday and starting back on Thursday. This also allows taking the weekend off. It seems easier to take on the mind-set of doing it every day, achieving your fitness goals faster, and then cutting back to a 3-day maintenance program. Now, I am trusting you to use common sense on what defines a workout. If you are playing one-on-one basketball every

Saturday morning with your teenager, then you can obviously count that as one of your workouts.

My fitness schedule after the birth of my son included the need for much fat burning, muscle building, and all-around body shaping. I started out walking 3 days a week. As soon as possible I moved to 4 days a week and began some light weight training on Tuesdays and Thursdays. I progressed to 5 days a week, which was my goal. Many times the demands of my schedule would not allow me to work out 5 times a week. Rather than consistently getting frustrated about not reaching my goals, I simply changed my goal to 4 times a week with a 5th workout as a bonus. In other words, I did everything I could to get a minimum of 4 workouts in per week. This included aerobics for at least 40 minutes twice a week and 20-30 minutes twice a week.

On my weight-training days, I started with the shorter aerobic sessions, then moved to weights. Working out 4-5 days a week eventually got me to my goals. Now, here is a very important key. My workouts were always progressive. Each week that I worked out, I worked out a little harder. As I mentioned earlier in the book, I started walking for 10 minutes. I progressed until I could walk 30-40 minutes 4 times a week. Then, I started walking 5 minutes, jogging 1 minute, walking 5 minutes, jogging 1 minute and so on. I gradually increased my jogging time. In several months, I was running a mile, walking a few minutes, running a mile, walking a few minutes, running a mile, and walking a few more minutes. I did this 2-3 days a week. I eventually worked my way up to running 3 miles straight on occasional workout days.

For my aerobics on my weight-training days, I used a machine called an elliptical cross-trainer. By cross-training with non-impact exercise, I was giving my knees and legs a break from jogging. I would do this for 20 minutes and then do my "muscle is a must" workout. After a few months progressing at one program, I would vary my workouts a little to keep my body changing. As my fitness level increased, my body got leaner and stronger. Some weeks I would do 3 days of weight training, and shorter, higher-intensity running. I would walk or jog at an easy speed for 60 seconds, then I would increase my speed for 60 seconds. I would repeat this pattern from 20-25 minutes. Did you ever have to run wind sprints in school? Well, this is very similar. Even though these workouts were shorter than my 45-minute aerobic workouts, because of the higher intensity, I was burning more calories in a shorter amount of time. I could also tell a tremendous metabolism boost was taking place. I am normally very cold natured, but I would stay warm for hours after high-intensity training.

It took my body several months to progress from overweight and out of shape back to an advanced fitness level. While it may take hard work to lose weight and get in shape, the rewards of fitness and health are more than worth it. One of my personal favorite benefits to being in really good shape is just feeling so healthy and strong. Because of so many years of struggling with losing weight, more than anything else, I think that I appreciate the freedom of not thinking so much about how I look or what I'm going to wear. It's just so wonderful when all of your clothes fit!

CHECKPOINT 9

Nutritional decisions

No matter how committed you are in the gym, it is imperative that you watch what you eat, especially when weight loss is your goal. It truly takes a commitment to better nutrition to help your body burn excess fat. I don't believe you should give up your favorite foods, but I do believe you should decrease the number of times you have them. Moderation is the key both to weight loss and weight maintenance.

Some say calories count; some say they don't. While you are deciding which "theory" to believe, your body is counting calories. Just as our heartbeat is an involuntary action, so is our body's ability to calculate calorie expenditure versus calorie intake. As tired as we get of hearing it, we must use more energy than we eat to burn stored fat. No, I am not recommending that you count every calorie. I am simply encouraging you to stay aware that your body does.

I so want you to achieve your goals and not experience the frustration that a hidden roadblock can cause. I have seen so many people who worked out hard in the gym, yet because of their refusal to temporarily give up certain foods, they thwarted their own progress. I have struggled long and hard with this myself. Suffice it to say that a combined effort of good nutrition and smart exercise is invaluable to our progress.

While a few extra pounds may not cheat us from many of the health benefits of exercise, if we have a past dealing with low self-esteem or poor body image, an extra 10 or 15 pounds can play havoc with our emotions. If we remain emotionally frustrated over a long enough period of time, it can be devastating to our goals and detrimental to our health. It is at this very point that we must decide to either change our strategy or to quit. I used to quit, take a break, and start the weight-loss cycle all over again. My idea of changing strategies was to find a new diet. Sometimes I would reach my goal weight and sometimes I wouldn't. When I did, it was "but for a moment," and then I would revert to my old habits and gain the weight again. I finally discovered the spiritual, mental, and physical balance required to defeat this issue once and for all.

As I progressively implemented the principles in this book, I changed. As I focused on spiritual growth and thinking about what God's Word says about me, I became so empowered spiritually and emotionally that I gained control over my eating and exercise habits. I was then equipped to successfully achieve my goals. I soon realized that an important part of enjoying a healthy lifestyle includes feeling good about who you are—spirit, soul, *and* body.

The power of agreement

What happens when we agree with God's Word? As I began to recognize good eating and exercise habits as agreeing with God's will for my life, I began to see results. I was no longer putting faith in fad diets or exercise gimmicks, but in God's promises. I realized that He had already promised me self-control, faithfulness, and endurance, and it was these spiritual strengths that would get me to my goals. I experienced firsthand how agreeing with and believing in His promises activated His power. This entire book is a testimony to what happened to me when I began agreeing with God's promises concerning weight loss, health, and fitness.

There is evidently another aspect to this power of agreement. Just as parents like it when their kids agree, so does God. There appears to be a multiplication of power, not just a doubling of power, when two or more people agree together in prayer. We see the principle of multiplication in Deuteronomy 32:30, where one can chase a thousand, and two can put ten thousand to flight. In Matthew 18:19-20 Jesus says, "That if two of you shall agree on earth as touching anything that they shall ask, it shall be done for them of my Father which is in heaven. For where two or three are gathered together in my name, there am I in the midst of them" (KJV).

It is really sad that I must take the time to explain that people who are pretending to be gathered together in the name of Christianity, yet are behaving contrary to God's law of love, are most assuredly not gathered in the name of Jesus. The name of Jesus represents the truth of God's Word from the book of Genesis to the book of Revelation and the principles of integrity that those scriptures uphold. For example, we are explicitly called to support our nation in prayer and in doing all that we can to support our military in defending the freedom and Godly rights that this country was founded on. We are not, however, called to "throw stones" or picket our next-door neighbors simply because they disagree with us. Christians are not even called to argue personal opinions, but simply to live a life of integrity, honor, and love. Our witness to others is not our "preach," but the integrity and faithfulness with which we live our lives.

It is with this understanding of the power of agreement and the integrity in the name of Jesus that I offer you my faith in agreement with yours. "Father, I ask in the name of Jesus that you will give this person wisdom in their commitment to change the quality of their life. I pray that if they have not yet experienced the unconditional love that you have for them, that they would take that step of faith to receive you as their Lord and Savior. Jesus, come into their heart and fill them to overflowing with your love and mercy. Show them continually what, when, and how much to eat, and give them wisdom concerning their exercise program. I ask you to help them find time to do all they need to do to attain their fitness goals. Father, we agree that as they spend time in your Word, you will speak to their heart and show them what to do. I release my faith now and agree with them that you will help them live a healthier, happier, and more prosperous life, in Jesus' name, Amen."

I believe with all my heart that this prayer is heard and that the power and blessings of God are released on your behalf. I want you to know that I will continue to believe with you and thank God for your progress. It has been an honor to share these truths with you. As always, remember that God is good, and it is His will for you to have a strong, healthy body!

A LITTLE KNOWN FACT...

Not only are exercisers more likely to maintain their weight loss, but in an article by Mark Lander of the Better Health and Medical Network, he states that reports have indicated that athletically and nutritionally fit individuals can be as many as 10-20 biological years younger than their chronological age.[9] Dr. Alex Leif, from the Harvard Medical School, has gone so far as to say that exercise is the closest thing to an anti-aging pill known.

ENDNOTES

1. "Fat Preference and Adherence to a Reduced-Fat Diet," from the Monell Chemical Senses Center, Philadelphia. Supported in part by PHS grant R01 DK 45294. Address reprint request to RD Mattes, Monell Chemical Senses Center, 3500 Market Street, Philadelphia, PA 19104-3308. Received June 3, 1992. Accepted for publication August 20, 1992.
2. *"Snack Attack,"* 13 July 2000, *Tip of the Day,* www.realage.com.
3. James Strong, *The New Strong's Exhaustive Concordance of the Bible* (Nashville: Thomas Nelson, 1990), p. 32.
4. Kenneth Cooper, *The Aerobics Program for Total Well-Being* (New York: Bantam, 1982), p. 26.
5. "Aging Fitness and Neurocognitive Function," Kramer AF, Hahn S, Cohen NJ, Banich MT, McAule Harrison CR, Chason J, Vakil E, Bardell L, Boileau L, Colcoombe A. 29 July 1999, National Library of Medicine, 27 September 2001.
6. "Stay Stronger Longer with Weight Training," *Harvard Health Letter*, 23.12 (1998): 1-3, editor—Leah R. Garnett.
7. Ibid.
8. "Influence of Leisure Time Physical Activity and Television Watching on Atherosclerosis Risk Factors in the NHLBI Family Heart Study," Kronenberg, F., Pereira, M. A., Schmitz, M.K., Arnett, D. K., Evenson, K. R., Crapo, R. O., Jensen, R. L., Burker, G. L., Sholinshy, P., Ellison, R. C., Hunt, S. C., *Atherosclerosis*, 2000 December, 153 (2):433-443, Real Age, Inc. Real Age Tip of the Day, Thursday, July 26, 2001.
9. "Turning Back Time," M. S. Lander, ETT, 1997 Better Health & Medical Network, 8 February 1999, www.betterhealth.com.

DATE:	WORKOUT:		
	SETS	REPS	WEIGHT
1. Standing Biceps Curls			
2. Military Press			
3. Side Lateral Raise			
4. Bent Over Row			
5. Triceps Kickback			
6. The Lunge			
7. The Squat			
8. Standing Straight-Toe Calf Raise			
9. The Push-up			
10. The Crunch			

DATE:	WORKOUT:	
	SETS	REPS
1. Crunch Variation #1		
2. Crunch Variation #2		
3. Crunch Variation #3		
4. Crunch Variation #4		
5. Crunch Variation #5		
6. Crunch Variation #6		
7. Crunch Variation #7		
8. Crunch Variation #8		
9. Crunch Variation #9		
10. Crunch Variation #10		

DATE:	WORKOUT:	
	SETS	REPS
1. Deep breathing		
2. Neck stretches		
3. Shoulder shrugs		
4. Desk pushers		
5. Fanny firmers		
6. Standing calf raises		
7. Plies		
8. Side leg lifts		
9. Hamstring stretches		
10. Lunges		

DATE:	WORKOUT:		
	SETS	REPS	WEIGHT
1. Standing Biceps Curls			
2. Military Press			
3. Side Lateral Raise			
4. Bent Over Row			
5. Triceps Kickback			
6. The Lunge			
7. The Squat			
8. Standing Straight-Toe Calf Raise			
9. The Push-up			
10. The Crunch			

DATE:	WORKOUT:	
	SETS	REPS
1. Crunch Variation #1		
2. Crunch Variation #2		
3. Crunch Variation #3		
4. Crunch Variation #4		
5. Crunch Variation #5		
6. Crunch Variation #6		
7. Crunch Variation #7		
8. Crunch Variation #8		
9. Crunch Variation #9		
10. Crunch Variation #10		

DATE:	WORKOUT:	
	SETS	REPS
1. Deep breathing		
2. Neck stretches		
3. Shoulder shrugs		
4. Desk pushers		
5. Fanny firmers		
6. Standing calf raises		
7. Plies		
8. Side leg lifts		
9. Hamstring stretches		
10. Lunges		

DATE:	WORKOUT:		
	SETS	REPS	WEIGHT
1. Standing Biceps Curls			
2. Military Press			
3. Side Lateral Raise			
4. Bent Over Row			
5. Triceps Kickback			
6. The Lunge			
7. The Squat			
8. Standing Straight-Toe Calf Raise			
9. The Push-up			
10. The Crunch			

DATE:	WORKOUT:	
	SETS	REPS
1. Crunch Variation #1		
2. Crunch Variation #2		
3. Crunch Variation #3		
4. Crunch Variation #4		
5. Crunch Variation #5		
6. Crunch Variation #6		
7. Crunch Variation #7		
8. Crunch Variation #8		
9. Crunch Variation #9		
10. Crunch Variation #10		

DATE:	WORKOUT:	
	SETS	REPS
1. Deep breathing		
2. Neck stretches		
3. Shoulder shrugs		
4. Desk pushers		
5. Fanny firmers		
6. Standing calf raises		
7. Plies		
8. Side leg lifts		
9. Hamstring stretches		
10. Lunges		

DATE:	WORKOUT:		
	SETS	REPS	WEIGHT
1. Standing Biceps Curls			
2. Military Press			
3. Side Lateral Raise			
4. Bent Over Row			
5. Triceps Kickback			
6. The Lunge			
7. The Squat			
8. Standing Straight-Toe Calf Raise			
9. The Push-up			
10. The Crunch			

DATE:	WORKOUT:	
	SETS	REPS
1. Crunch Variation #1		
2. Crunch Variation #2		
3. Crunch Variation #3		
4. Crunch Variation #4		
5. Crunch Variation #5		
6. Crunch Variation #6		
7. Crunch Variation #7		
8. Crunch Variation #8		
9. Crunch Variation #9		
10. Crunch Variation #10		

DATE:	WORKOUT:	
	SETS	REPS
1. Deep breathing		
2. Neck stretches		
3. Shoulder shrugs		
4. Desk pushers		
5. Fanny firmers		
6. Standing calf raises		
7. Plies		
8. Side leg lifts		
9. Hamstring stretches		
10. Lunges		

DATE:	WORKOUT:		
	SETS	REPS	WEIGHT
1. Standing Biceps Curls			
2. Military Press			
3. Side Lateral Raise			
4. Bent Over Row			
5. Triceps Kickback			
6. The Lunge			
7. The Squat			
8. Standing Straight-Toe Calf Raise			
9. The Push-up			
10. The Crunch			

DATE:	WORKOUT:	
	SETS	REPS
1. Crunch Variation #1		
2. Crunch Variation #2		
3. Crunch Variation #3		
4. Crunch Variation #4		
5. Crunch Variation #5		
6. Crunch Variation #6		
7. Crunch Variation #7		
8. Crunch Variation #8		
9. Crunch Variation #9		
10. Crunch Variation #10		

DATE:	WORKOUT:	
	SETS	REPS
1. Deep breathing		
2. Neck stretches		
3. Shoulder shrugs		
4. Desk pushers		
5. Fanny firmers		
6. Standing calf raises		
7. Plies		
8. Side leg lifts		
9. Hamstring stretches		
10. Lunges		

DATE:	WORKOUT:		
	SETS	REPS	WEIGHT
1. Standing Biceps Curls			
2. Military Press			
3. Side Lateral Raise			
4. Bent Over Row			
5. Triceps Kickback			
6. The Lunge			
7. The Squat			
8. Standing Straight-Toe Calf Raise			
9. The Push-up			
10. The Crunch			

DATE:	WORKOUT:	
	SETS	REPS
1. Crunch Variation #1		
2. Crunch Variation #2		
3. Crunch Variation #3		
4. Crunch Variation #4		
5. Crunch Variation #5		
6. Crunch Variation #6		
7. Crunch Variation #7		
8. Crunch Variation #8		
9. Crunch Variation #9		
10. Crunch Variation #10		

DATE:	WORKOUT:	
	SETS	REPS
1. Deep breathing		
2. Neck stretches		
3. Shoulder shrugs		
4. Desk pushers		
5. Fanny firmers		
6. Standing calf raises		
7. Plies		
8. Side leg lifts		
9. Hamstring stretches		
10. Lunges		

DATE:	WORKOUT:		
	SETS	REPS	WEIGHT
1. Standing Biceps Curls			
2. Military Press			
3. Side Lateral Raise			
4. Bent Over Row			
5. Triceps Kickback			
6. The Lunge			
7. The Squat			
8. Standing Straight-Toe Calf Raise			
9. The Push-up			
10. The Crunch			

DATE:	WORKOUT:	
	SETS	REPS
1. Crunch Variation #1		
2. Crunch Variation #2		
3. Crunch Variation #3		
4. Crunch Variation #4		
5. Crunch Variation #5		
6. Crunch Variation #6		
7. Crunch Variation #7		
8. Crunch Variation #8		
9. Crunch Variation #9		
10. Crunch Variation #10		

DATE:	WORKOUT:	
	SETS	REPS
1. Deep breathing		
2. Neck stretches		
3. Shoulder shrugs		
4. Desk pushers		
5. Fanny firmers		
6. Standing calf raises		
7. Plies		
8. Side leg lifts		
9. Hamstring stretches		
10. Lunges		

COME AND JOIN THE PRAYER TEAM!

Marty Copeland would like to invite you to join her prayer team of over 20,000 strong and growing!

Let Marty pray for you and join your faith with hers as you develop a lifestyle of faith, freedom and self-control. The Higher Fitness Prayer Team is rapidly increasing as people are enjoying the opportunity of giving and receiving the blessing of daily prayer.

To sign up, just go to MartyCopeland.com. You will receive a new prayer each month with an encouraging word from Marty, a short prayer, scriptures to stand on and a *Higher Fitness* challenge. God is answering prayers and changing lives through the power of agreement. There is no obligation other than your commitment to pray. Your life will never be the same again!

You will receive rich rewards as you sow in prayer toward the success of other team members because Galatians 6:7 (KJV) says, "Be not deceived; God is not mocked: for whatsoever a man soweth, that shall he also reap."

For health and fitness information, encouragement and spiritual support, go to Marty's *Higher Fitness* Web site at www.martycopeland.com.

OTHER PRODUCTS BY MARTY COPELAND

Walkin' Strong!

Level One: Two 30-Minute Beginner Workouts, CD includes easy instructions to customize your workout; information on walking and safety tips; motivational instruction; powerful encouragement and faith-filled prayer.

Walkin' Stronger!

Level Two/Three: 45 and 60-Minute Intermediate/Advanced Walking Workouts, CD includes easy instructions, information, encouragement and faith-filled prayer.

Totally Strong!

Level Three: 60 Minutes of Extreme Workout Music, CD includes music at 140 BPM. Great for the gym, running, jogging, power walking, aerobic classes or anytime!

Arise and Walk

God's Word coupled with a walking workout. Includes 60-minute video, 60-minute music cassette, instruction card and prayer card.

Health & Fitness as It Pertains to Life & Godliness

No matter how many times you've failed at dieting, God created you for victory and success. (4-tape series)

Present Your Body!

Take the first step toward beating the weight-gain, weight-loss cycle. Present your body to God and yield to His Anointing. (single tape)

Hannah's Miracle Child *(NEW!)*

Marty knew God is no respecter of persons. If He opened Hannah's womb and she conceived, He could do the same for her. That's all she needed to get on the road to believing God for her own children. Whatever you're believing God for, don't give up. Set your faith to have the same stance as Hannah, and *don't* let go until you get what is rightfully yours! (single CD)